TRIUMPH OF THE LAITY

TRIUMPH OF THE LAITY

Scots-Irish Piety and the Great Awakening, 1625–1760

Marilyn J. Westerkamp

New York   Oxford
OXFORD UNIVERSITY PRESS
1988

Oxford University Press

Oxford   New York   Toronto
Delhi   Bombay   Calcutta   Madras   Karachi
Petaling Jaya   Singapore   Hong Kong   Tokyo
Nairobi   Dar es Salaam   Cape Town
Melbourne   Auckland
and associated companies in
Beirut   Berlin   Ibadan   Nicosia

Library of Congress Cataloging-in-Publication Data

Westerkamp, Marilyn J.
  Triumph of the laity.

  Bibliography: p.
  Includes index.
  1. Revivals—Middle Atlantic States—History—
17th century. 2. Revivals—Middle Atlantic States—
History—18th century. 3. Great Awakening. 4. Scots
Irish—Middle Atlantic States—History—17th century.
5. Presbyterian Church—Middle Atlantic States—History—
17th century. 6. Scots Irish—Middle Atlantic States—
History—18th century. 7. Presbyterian Church—Middle
Atlantic States—History—18th century. 8. Middle
Atlantic States—Church history. 9. Middle Atlantic
States—History—Colonial period, ca. 1600-1775.
I. Title.
BR520.W47   1988   277.4'07   87-1673
ISBN 0-19-504401-0

9 8 7 6 5 4 3 2 1

Printed in the United States of America
on acid-free paper

For Cynthia Montague

Acknowledgments

Triumph of the Laity was first prepared as a dissertation in the Department of American Civilization at the University of Pennsylvania. While researching and writing the dissertation, and then revising the manuscript for publication, I have become indebted to many individuals and institutions. To Murray G. Murphey, teacher and friend, I am deeply grateful for his stimulating instruction and continuous intellectual and moral support. Though I cannot begin to describe the structure he has provided for my ideas or the contributions he has made to my research, it may be enough to say that for over a year he pertinaciously and perspicaciously asked, "When are you going to Ireland?"

I must also thank Richard S. Dunn for the advice and support he has given me since I first entered graduate school. From the time I began searching the archives until the final draft of my dissertation went through committee, I have received invaluable criticism from him. Many of the revisions for publication—particularly the addition of a seventh chapter—reflect his perceptions and recommendations. For the expertise he brings to historical research and the care he takes with a student I am especially grateful.

My travels through Ireland, Scotland, and England were made possible by a Penfield Travel Grant from the University of Pennsylvania. Michael Seymour and Wolfson College arranged my stay in Cambridge, England, and Dr. Ronald Buchanan, director of the Institute for Irish Studies, The Queen's University, Belfast, provided me with an institutional home during my stay in Northern Ireland. Many people in Britain helped make my stay pleasant and my research productive. To John M. Barkley, David Hempton, Finley Holmes, and Eric McKimmon I owe thanks for long, exploratory conversations and bibliographic and archival assistance. I was permitted the use of the Public Records Office of Northern Ireland

viii   *Acknowledgments*

and the Presbyterian Historical Society of Ireland in Belfast. Similarly, the Scottish Records Office and the National Library of Scotland in Edinburgh opened their doors to me. I express thanks to the staffs of all four archives.

I am most indebted to the Gamble Library, Union Theological College, Belfast. There I spent most of my research time, and there I met staff librarian Anne McConnell, who gave me unlimited assistance and offered her friendship. Without a doubt, her support and advice doubled the productivity and ultimate usefulness of my research time.

Friends and institutions on this side of the Atlantic also provided assistance and support. Grants from the Department of American Civilization at the University of Pennsylvania and from the Philadelphia Center for Early American Studies enabled me to complete my research. Many other individuals have shared information with me or have patiently discussed my research and ideas. For their generosity I thank Dennis Barone, Richard R. Beeman, Drew G. Faust, William J. Frost, Melvyn Hammarberg, Patricia Hanen, Guy S. Klett, Bruce Kuklick, Ned Landsman, Michael McGiffert, Eleanor McLaughlin, Janice Radway, Jean Soderlund, Mariana Wokeck, and Michael Zuckerman. At this point I also wish to thank William B. Miller, Gerald G. Gillette, Mary Plummer, and the staff of the Presbyterian Historical Society in Philadelphia for their assistance.

The revision process is generally tedious, but the gracious and skilled assistance I have received has made my life infinitely easier. Cynthia Read and the staff of Oxford University Press have been helpful and supportive throughout. I especially appreciate the care they have taken through the evaluation and editorial process and the commitment they have demonstrated to the success of this work. The detailed attention of the anonymous evaluator and the impressive skill of the copy editor, Jennifer Quinn, have significantly strengthened the book. At Clarion University I was fortunate to have the assistance of the clerical staff of the history department, including Sue Lemmon and Debra Marchand. The staff of the Clarion Library, especially Judy Bowser of Interlibrary Loan, has been particularly helpful. When appropriate maps could not be found, my sister, Kit Westerkamp, provided them. To Marcy Leavy, master of the word processor, who typed the entire manuscript and entered the revisions—sometimes two or three times over—I express my appreciation of her skill, efficiency, and infinite good humor.

Finally, I want in some way to acknowledge those intangibles reflected not in the text and notes but in the very fact that the book was finished. I have been at Clarion University only a short while, and yet my colleagues here—Steven Piott, Suzanne Van Meter, and Brian Dunn—have provided the nurturance and enthusiasm of lifelong friends. Others at great distances have frequently shared my joys and frustrations. To Carl Drott, Sybil Lipschultz, Mary McConaghy, Maggie Morris, Catherine Robert, Eva Thury, and Janet Tighe I hope that I have reciprocated and will return as they have given. Lastly, I have had the care and encouragement of my parents from the earliest days of my education. Nothing has been more important to this book than the love and support of family and friends.

Contents

TRIUMPH OF THE LAITY

Introduction

All students of American history have been taught that the Great Awakening was one of the most significant events of the eighteenth century. The Great Awakening has been considered a kind of chronological watershed, dividing the colonial period from the revolutionary period. The historian, heretofore preoccupied with the political, economic, social, and intellectual developments of distinct colonies or regions, has been told to see post-1740 British North America as a single unit: provincial America. To the Great Awakening have been attributed the massive upheavals and changes of the century, not only within church organizations but in the very infrastructure of society. Thus has the awakening been used to explain major changes observed in the eighteenth century, primarily the American Revolution. Yet despite the immense importance that the Great Awakening has assumed for the historical picture, no one has produced a single, comprehensive, synthetic work that attempts to pull together, successfully or otherwise, all the facets of the movement into a united, coherent interpretation. Instead, the canonical view was formulated in bits and pieces by scholars working within a specific region until a common set of assumptions was recognized by all, only to be challenged as continued research proved inconsistent with the canon.

The first portion of the canon is definitional: What was the Great Awakening, when did it occur, and where? The general response has been that the Great Awakening was a brief, intense period of religious revival that "swept through the American colonies between 1739 and 1742."[1] Jon Butler's witty attempt to turn this scholarship upside down was amply justified.[2] Several regions in the colonies remained untouched by revival throughout the eighteenth century. For example, Stephanie Wolf's study of Germantown shows the residents in those communities surrounding Philadelphia relatively unaffected by a movement that supposedly rocked

3

Pennsylvania. Other scholarship has demonstrated that the revival never quite reached South Carolina.[3] Perhaps the reason for this geographic misperception is that most studies of the awakening have been focused upon New England, which was indeed swept by the revival.

These same New England histories challenge the chronological boundaries assigned to the movement, for all acknowledge the official beginning of the Great Awakening in the revival at Northampton in 1734. In fact, in western Massachusetts ministers had been having small revival-type success at least fifteen years prior to the Northampton revivals, with Solomon Stoddard enjoying "harvests of souls" as early as the 1680s.[4] The general description further erred in determining the date that the awakening ended; for, as Rhys Isaac has pointed out in his recent articles and book, the revival did not really affect Virginia until the late 1750s.[5] Still, historians should not reject "Great Awakening" as a descriptive label. They must merely recognize that this was not an event but an eighteenth-century movement; that there were many revivals, not one, that sprang up in various regions such as New England, Pennsylvania, and the Chesapeake; and that the narrowest chronological boundaries would probably be 1725 to 1765.

Another aspect of the standard interpretation involves the effects of the Great Awakening upon colonial society. The revival has been seen as an innovative force, transforming cities and the countryside, uniting disparate peoples in preparation for the American Revolution. Alan Heimert outlined this position so effectively that few have since challenged him. In his *Religion and the American Mind,* Heimert demonstrated that the awakening's evangelical/Calvinist theology was a primary influence upon revolutionary and, later, Jacksonian ideology.[6] From a completely different standpoint Harry Stout and Rhys Isaac have argued that, intellectual influences aside, the awakening provided new sets of symbols and methods of communication by which the revolutionary ideology was spread throughout New England and Virginia.[7] Yet the problems with these two interpretations are highlighted by the reasonableness of recent refutations. Nathan Hatch claimed that the most important religious aspects of this political ideology were those beliefs shared by both Old and New Lights, and Gary Nash could argue a connection between politics and religious upheaval only for Boston.[8] Instead of trying to evaluate the extent of the awakening's effect upon society, perhaps the causal model should be turned on its side. Rather than causing the Revolution, the Great Awakening may have resulted from the same societal changes that resulted in the Revolution. In other words, the challenges of the eighteenth century first produced widespread religious revivalism and then a political revolution.

This construct would account for the apparent fit or incongruity in applying the Great Awakening to the American Revolution model.

The natural next step would be to suggest the causes of the awakening, and the canon does, in fact, include two possibilities. The first causal model was essentially the product of one mind—Perry Miller—and it has been adopted, with minor alterations, by intellectual historians ever since. Enamored of the brilliance of Jonathan Edwards, Miller and his followers examined the theology of Edwards as a reworking of old, orthodox Calvinism to meet the challenges of the new, scientific Age of Enlightenment. This is particularly apt since Edwards was also the identified leader of the awakening in New England and produced the best theological apologia for revival.[9] The obvious flaw in this explanation is that, except among the educated clergy, Edwards was a regional figure. Unlike George Whitefield or even Gilbert Tennent, Edwards did not preach outside the bounds of New England. Moreover, this model implies that the people as a whole understood science and theology well enough to be uncomfortable with a recognized conflict between the old religion and the new science and therefore embraced a new, synthetic ideology. In fact, this model implies a shared ideology among all the revival participants. The relative sophistication of the average lay person aside, it is demonstrable that the participants did not share a theology. Unless one is willing to remove Virginia, North Carolina, and many of the middle-colony communities from the scope of the Great Awakening, scholars are left with an ideologically diverse revival whose participants accepted a spectrum of theologies, from conservative Calvinism to Methodist Arminianism.

An equally unsatisfactory alternate explanation is the social unrest or class conflict model. This interpretation views the Great Awakening as the convergence of the disaffected lower classes against the elite. In an early exposition of this thesis, Martin Lodge posited anticlericalism as a primary force driving the community toward awakening in the middle colonies. The problem with his argument, however, is his dependence upon the colonial environment providing new challenges to which immigrants responded. He hardly considers the possibility that clerical-lay tensions may have had deep roots in the Old World, and he neglects the fact that the religious environment in the middle colonies was very much like the one left behind by large numbers of immigrants from Ireland. As the new social history progressed, this hypothesis and its investigation produced some remarkable and fascinating historical scholarship. Unfortunately, linked as it is to in-depth, regional investigation, the results of this research have been contradictory. James Walsh and Gerald Moran both found that class conflict had very little impact upon the revival in two Connecticut

towns; Rhys Isaac discovered in the Great Awakening an organizing principle for lower-class protest; and Gary Nash found evidence in his examination of colonial cities to support this conclusion for Boston and question
its applicability to New York and Philadelphia.[10]

In addition to the problem of regional variations, the social conflict
hypothesis suffers from a basic misunderstanding of the revival. The Great
Awakening was not primarily a social but a religious movement. To deny
or minimize the religious components is to simplify greatly the structure of
the phenomenon and to remove the revival from the circumstances within
which the contemporary participants experienced it. Using the construct of
a religious system as the starting point for an investigation of the awakening significantly broadens the field of inquiry by connecting aspects of
eighteenth-century culture that were otherwise unconnected—for example,
elite philosophy and popular behavior. Furthermore, this seems the most
historically appropriate approach since it places the revival into its original
contemporary context and provides a code through which the historical
evidence can be sorted and analyzed.[11]

What is meant, then, by a religious system? As a working framework
for the seventeenth and eighteenth centuries, I propose a definition that
includes both substantive and functional aspects. Briefly stated, a religious
system is a conceptual system concerned with God, humanity, the supernatural, and the natural world. These ideas are discussed in a consciously
constructed language and theology and are acted out through symbolic,
communal behavior patterns or rituals that bring emotional satisfaction to
the participants.

This definition is based upon a model proposed by Clifford Geertz in
his "Religion as a Cultural System."[12] I have appropriated Geertz's
emphasis upon the symbolic quality of the system and his discussion of
religion as fulfilling a specific function. The restriction concerning content
is my addition to the analytical schema; it is actually a historical conve-
nience rather than a universal characteristic. In the twentieth century, for
example, secular ideologies might well be religions in the symbolic and
functional sense, even though a rejection of God and the supernatural may
be part of the conceptual framework. In other words, I have accepted the
seventeenth- and eighteenth-century substantive definition of religion; the
anthropological definition will inform my analysis.

Restricting the definition of religion to belief systems about a deity is
not meant to limit historical application, for example, by excluding scientific beliefs from such a system. Rather, the restriction describes the way in
which such beliefs would be implicated within the conceptual schema.
Since this content is generally expressed in behavioral as well as linguistic

symbols, the study of a religion must include an analysis of its symbol systems. Finally, the purpose of religion—to bring emotional satisfaction or, in eighteenth-century terms, spiritual fulfillment—might be explained as providing meaning for the logically meaningless. Examples generally cited to elucidate this functional concept include the problem of pain (Why and how should a person suffer?) or the problem of evil (Why do the good suffer and the wicked prosper?). The religion of seventeenth-century Scotland, for instance, developed certain rituals that helped people cope with a level of political chaos that might otherwise have destroyed their communities.[13]

This definition sets up an analytical structure that divides religious systems into four distinct, interdependent components: shared beliefs, common rituals, institutional manifestations, and participants. The ideological component includes both the theology developed self-consciously and debated by the intellectual elite and the popular belief system that would be gleaned by the nonelite from the original theology. Rituals are symbolic behavior patterns established and/or reinforced by the community. They include both community events and private behaviors that all or most members of a community act out individually. For example, private prayer would not necessarily be a common ritual, but Puritan devotional exercises as structured in seventeenth-century New England would be considered so since the individual participants followed privately a behavioral formula laid out in the prescriptive literature. The institutional component, of course, includes the churches themselves, their policies, organizations, record keeping, and procedures devoted to the maintenance of the institution. This would not only encompass the judicatory system in the Presbyterian church but also the official documents, the seminaries, and their relations with the civil government. The final component, frequently forgotten by religious historians, includes the participants themselves, both the leaders and their followers—what might be termed the social component.

What must be understood is that each of these components develops apart from, yet is influenced by, the others. That is, each component has a separate existence that transcends the realities of the immediate system, yet none can exist without the others. Consider a theology, such as orthodox Calvinism. There is a systematic compilation of beliefs recognized by theologians as Calvinism. This theology was developed by individual thinkers, and key features have been reinforced by certain rituals; but Calvinist theology has also been influenced by other intellectual systems, and it is frequently defined through comparisons with ideological alternatives. Were there no orthodox Calvinist believers today, historians could still discuss an identifiable theology. The intellectual system exists independently of its believers, yet the pattern of its development has been determined by those very believers and their behavior.

Obviously the same could be said for the people. Their behavior has been at least partly determined by their beliefs; yet many other factors—economic, demographic, political—affect people's circumstances and choices. Individual or community adaptation to historical situation has frequently created, challenged, or modified a group's belief system. And few people would challenge the independent existence of a church establishment, although the church obviously could not exist without members. It is the independence of ritual that may be the most difficult concept to comprehend. However, the celebration of Christmas in the twentieth-century United States amply demonstrates the potential strength of ritual for its own sake. Many Americans are indifferent to the theological background of Christmas; others are nonbelievers. Yet neither ideological indifference nor rejection interferes with family and community celebrations. Trees are decorated and gifts exchanged because it is Christmas; the ritual itself has become as important to the people (if not more so) as the beliefs honored by the celebration.

By investigating religious phenomena in terms of these four components, primary or determinant features can be distinguished from peripheral characteristics. That is, a structure has been established within which religious phenomena can be explicitly defined and systematically compared. In fact, this model applied to the Great Awakening provides both an explanation for the limited success of recent historical research and an alternate research proposal.

Intellectual historians have focused upon the theology surrounding the Great Awakening. While some scholars have claimed the predominance of a monolithic ideology, their combined efforts prove that this was not the case. Whatever their respective believers hold in common, orthodox Calvinism and Methodist Arminianism are ideological opposites. Adherents to both systems participated in the awakening, indicating that ideology was a secondary characteristic of the movement. The social historians have discovered the same discrepancies within their own methodological arena—the study of people. In some regions economic status is critical; in others it hardly seems to matter. The awakening thus cannot be described across the colonies as primarily a social protest within a religious structure. Nor can the awakening be described as an institutional phenomenon, since some churches, such as the Baptist church, wholeheartedly supported it, some groups, such as the Quakers, absolutely rejected the revival, and others, notably Congregationalists and Presbyterians, sported both advocates and detractors.

Clearly what remains as the defining characteristic of the Great Awakening is ritual; and, indeed, whenever scholars discuss the awakening, their investigative environment is, generally, the ritual behavior.

> Flamboyant and highly emotional preaching made its first widespread appearance in the Puritan churches . . . and under its impact there was a great increase in the number and intensity of bodily effects of conversion—fainting, weeping, shrieking, etc.[14]

This coincides nicely with eighteenth-century perceptions of the event. Contemporary divines were so awed by the extent and intensity of the "great and general awakening" that, for several months, they published a periodical entitled *Christian History* that comprised a series of reports on the great works of religion throughout the colonies and Great Britain. Of course, the eighteenth-century writers explained such phenomena as, simply, the work of the Holy Spirit. Twentieth-century scholars, unwilling to drop the question so easily, still begin with descriptive quotations or allusions to this frenetic behavior; then they proceed to discover its cause through its relationship to some other factor, be it class protest or philosophical developments. Returning to the Great Awakening as a phenomenon characterized by its rituals, perhaps another sort of question should be explored: Could these rituals have a life of their own?

The history of these rituals has been the focus of my research. Beginning with the eighteenth-century revivals themselves, I have observed their structure and quality, the definitive characteristics and peripheral details. Deliberately, and sometimes with difficulty, placing ideas to one side, I have sought the origin of these rituals as behavior patterns. Participants have been examined in relation to their symbolic behavior, community identity examined in terms of institutional religion. When ideological issues were addressed, it was with an eye toward their role in the symbolic network rather than the quality and content of the philosophy.

This, then, is a history of the rituals that characterized the Great Awakening, rituals that I will identify as revivalism. Although various revivalist services will be described throughout, I would like, at the beginning, to establish some key features of rituals I would classify as revivalist. The first is size, or number of participants; revivalist meetings are not necessarily large on an absolute scale, but they must be larger than normal religious services. Many revivals had thousands attending, but considering the wide geographic dispersion of the residents of eighteenth-century rural

America, an attendance of several hundred at a gathering would be quite large enough. A second feature is the duration of the meeting. Most religious services of this period lasted, at most, three to four hours; a revival meeting might easily last three to four days. The most remarkable common characteristic is the intense emotional response, often manifested in bizarre physical symptoms, to leaders who by rhetoric and style deliberately provoked such response. Finally, the purpose of the ritual, one frequently but not always recognized by the participants themselves, is to free participants from guilt, shame, and sin and enable them to identify with a community that is loving, good, and protected by God—in the revivalists' own terms, to experience conversion. Participation in the ritual allows participants to attain or retain membership in the desired community, while the community itself is elevated to a holier status.

One other methodological problem that should be considered is the twentieth-century scholars' frequent equation of revivalism with revitalization. As first laid out by Anthony Wallace, revitalization describes communities undergoing significant renewal. A community experiencing a slow institutional or cultural decay may suddenly be caught up in the activities of a charismatic leader or contingent of leaders. By infusing old traditions with new ideological structures or developing and adapting rituals to changing circumstances, leaders can revitalize their communities, reverse the decline, and, with amazing speed, strengthen the people and their cultural integrity.[15] Revivalism may be a part of a revitalization movement, but is not necessarily so. In the early twentieth century, for example, revivalism was a conservative force working to maintain the status quo.

Many historians, however, have argued that the Great Awakening was a revitalization movement, providing the decadent New England religious culture with new life and vitality. This interpretation includes an acceptance of Perry Miller's theory that from 1660 onward New England was suffering a decline of piety and that the society as a whole was desperately in need of and prepared for renewal.[16] However, if Miller's thesis is rejected, the revitalization argument is not so strong. A more critical point involves the recurring New England focus. Whether or not Puritan New England, by virtue of its age, was in decline, the remaining colonies, except for Virginia, were in the midst of feverish growth. The concept of revitalization assumes the existence of an established, traditional society that is becoming stale. A culture that has yet to mature and stabilize cannot be "revitalized." In light of historical knowledge about eighteenth-century culture, a far more reasonable conclusion would identify the Great Awakening as part of this cultural growth rather than as "revitalization."

William Howland Kenney has argued that the revitalization argument does apply to the middle colonies. Using both Wallace's scholarship and Max Weber's theory of charisma, Kenney posits Whitefield in the charismatic leadership role. His hypothesis is founded upon the assumptions that most churches in the middle colonies were in decline, and that the people were therefore seeking salvation. While his argument may or may not be true for the Anglican community, I find I must reject it for the Presbyterian church. The colonial church was growing at a quick pace, uniting diverse ethnic groups under a single organizational structure and taking steps to provide ministers and services for the people. I must also reject what I consider an artificial division between religious culture and other aspects of life. Wallace's description of the Iroquois cultural disintegration, for example, points out that every aspect of society—social structure, economic supports, military strength as well as values and prestige—was decaying concurrently. If the community was strengthening its economical, political, and social structures, why would problems within the institutional churches countermand that progress and create a need for revitalization? The newly arrived immigrants were not a demoralized community, but a community on the brink of success.[17]

Thus I am exploring the origins and progress of the awakening within a changing—and positively changing—society. In trying to explore both the intense emotionality and the social complexity of the awakening, I have first isolated the rituals, and then placed them back into the cultural environment. I began with the Presbyterian community in Philadelphia, a single community in the middle colonies violently split by the events of the 1730s and 1740s. Here I used congregational administrative records and membership lists in conjunction with other social documentation, such as tax lists and probate records. The goal was not only to discover one institutional environment that nurtured the awakening, but to identify as many participants as possible in order to discuss their actual behavior. To further increase my understanding of the progress of the revival in Philadelphia, I depended upon the more impressionistic reminiscences, journals, and correspondence. Finally, I turned to sermons and polemical pamphlets as both historical record and ritual artifact.

I grew very dissatisfied with the limited answers provided by a single-community study, so I broadened my search to include records of the colonial Presbyterian church, actually including, during most of the eighteenth century, only congregations in the middle colonies—southern New

York, New Jersey, Pennsylvania, Delaware, and a few organizations in the
Chesapeake area. Again, I not only used synod and presbytery records, but
I enlarged the pool of qualitative evidence with the papers and publications
of Presbyterian leaders in these other regions. Pamphlets were especially
useful in filling the two-, six-, or fifteen-year gaps found in various sets of
administrative records. To round out this survey of the middle colony
region, I looked at materials concerning the regional educational institutions, such as William Tennent's seminary, or the "Log College," and the
College of New Jersey, and the progress of academies and colleges in
Philadelphia. Finally, I turned to the sermons and records of the major
itinerants, especially Whitefield, since they experienced high success in
this area.

My research was limited geographically to the middle colony region
and institutionally to the Presbyterian church, both the result of an initial
focus upon the enthusiastic Philadelphia response to George Whitefield.
The geographic boundaries have kept the investigation of the Great Awakening under control; the use of the Presbyterian church has made available
a vast amount of data. The constant risk has been that this book might
become a history of the Presbyterian church. The problems faced by the
new church organization were fascinating and distracting; but I discovered
that with the centrality of the Great Awakening for eighteenth-century
Presbyterians, the major conflicts experienced by the growing communion
were all related indirectly, if not directly, to the divisions between supporters and opponents of the revival.

Expanding the research program to include the entire Presbyterian
network in the middle colonies increased the number and scope of the
questions without really providing any more answers. During the first forty
years of the church's history (1705–45), several battles were fought over a
variety of issues—creeds, ministerial education, George Whitefield—and
each time patterns of alliance were somewhat different. The available data
could explain the behavior of any number of individual clerics involved
with each issue. Variables such as childhood residence, regional history
and location of congregation, educational background, friendships, and
family connections all came together to determine different responses to
the same challenges. Certainly no patterns developed, and very little of the
information offered any explanation for the widespread revivalism in the
area.

The failure to discover any clear patterns related to revivalism led me
to alter my approach in two ways. First, I used the American sources to
piece together lay behavior, especially those aspects that were somewhat
independent of the clerical leadership. Obviously the rituals involved many

more lay persons than ministers, and my almost involuntary emphasis upon the ministers had come from the kind of evidence available. However, as the problems of this focus became apparent, I decided to try to uncover popular intention and action. The second change resulted from the discovery that most Presbyterians in this region, and an increasing proportion as the century progressed, had emigrated from Ireland. I therefore went to Northern Ireland and Scotland, where I examined a diverse spectrum of institutional and individual records related to the Protestant communities during the seventeenth and eighteenth centuries. I continued my efforts at identifying lay as opposed to clerical behavior and found, as might have been expected, that many of the observed developmental patterns of the eighteenth century grew out of the existence of separate spheres for ministerial and lay activity and the interdependence and mutual influences of the two groups.

All of these considerations gave the final project an enormous scope. Research covering three regions—Scotland, Ireland, and the middle colonies—over 140 years is not easily synthesized or controlled. The analysis focused upon only one cultural development, revivalism, but even keeping to that restriction was made difficult by the social complexity of the period. After all, this revivalism must be presented within its historical context, for this context provides a code by which the symbolic language and behavior can be translated. Moreover, since religious systems are not self-contained but frequently influenced by factors outside the control of the system, a knowledge of the general history of the period is needed to construct a complete explanation of one particular cultural phenomenon. Thus I have tried to balance the in-depth description and analysis of revivalism against brief coverage of those political, economic, and social factors that greatly affected the religious tradition.

The final answers gleaned from my research are historical ones, going back to 1625. Therefore, although my research began with the Great Awakening, the story I present begins in the seventeenth century with the plantation of Northern Ireland. The chapters are arranged chronologically, documenting the establishment of certain religious traditions in Ireland and Scotland in the early seventeenth century, the continuation of those traditions through the following hundred years, and their migration to the middle colonies with the Scots-Irish people. As the narrative progresses to the eighteenth century, the awakening is examined as an interaction between the traditional religious behavior and a new cultural environment. This interaction is accentuated by a comparison of the awakening in the middle colonies with the modest and peculiar progress of revivalism in eighteenth century Ireland and Scotland. Not only does the comparison point up the remarkable quality of the colonial phenomenon, but it serves to identify and distinguish causal and correlative factors in a way that could not be accomplished by the conventional, colony-specific approach.

The Great Awakening in the middle colonies represented neither innovative religious behavior nor a statement of challenge to the establishment. Rather, that revivalism, first observed in the colonies during this time, was actually part of the Scots-Irish religiosity, a tradition that flourished under the encouragement afforded by the colonial ministers. Its importance was attested by the jealousy with which the Scots-Irish guarded their revivalism. It cannot be coincidental that outbreaks of revivalism in Pennsylvania and New Jersey followed directly after large-scale migration from Ireland, nor that the awakening spread to Virginia and the Carolina back country during the years when the Scots-Irish moved south. Presbyterianism might be exchanged for Methodism, and Calvinism for Arminianism, before the revival was forsaken. It had originated long before Whitefield toured the colonies, and continued long after the last itinerant preacher had died. From this vantage point, the Great Awakening was one episode, albeit a magnificent one, in the life and growth of a religious tradition.

1

Genesis of Revival

> I have seen them myself stricken, and swoon with the Word—
> yea, a dozen in one day carried out of doors as dead, so
> marvellous was the power of God smiting their hearts for sin,
> condemning and killing; and some of those were none of the
> weaker sex or spirit, but indeed some of the boldest spirits . . .
> the stubborn, who sinned and gloried in it, because they feared
> not man, are now patterns of sobriety, fearing to sin because
> they fear God; and this spread through the country to admira-
> tion, so that, in a manner, as many as came to hear the word of
> God, went away slain with the words of his mouth . . .[1]

Here is yet another description of stout-hearted converts swept up by the
popular religious revivalism that characterized the Great Awakening during the middle decades of the eighteenth century. Yet this account was
written by a minister who had never been to America, written almost one
hundred years before George Whitefield toured the colonies. The writer
was Andrew Stewart, a seventeenth-century Irish Presbyterian minister; he
was recording events that occurred when he was a child of eight or nine in
the province of Ulster.

Ulster was peculiar compared to the rest of the British Isles. It had
been the stronghold of the powerful clans Tyrone and Tyrconnel, and the
last Irish province to have withstood all English attempts to conquer and
gratuitously civilize its barbaric inhabitants. When James VI of Scotland
ascended the English throne, he inherited the Irish problem and, specifically, the Ulster problem. Following the example set by the Tudors, his
solution was colonization, but with this difference: he used Scottish colo-

nists. Since his colonists planned to stay, they brought their own ministers. Although James VI/I established the Anglican church as the Protestant Church of Ireland, several of its bishops and clerics were Scots grounded in a solid tradition of Knoxian austerity. In 1625, less than twenty-five years after the Scots had begun to settle Ulster in sizeable numbers and less than ten years after the first Scottish missionaries had arrived, Ulster was shaken with the "Six-Mile-Water Revival."

Although the initial outbreaks clustered around the Six-Mile-Water area, close to the town of Antrim, it eventually spread throughout Counties Down and Antrim. The revival lasted for eight years—until 1633, when Charles I and Archbishop William Laud began to restrict the movements of the dissenting ministers through suspensions and depositions. Recognized by Irish church historians as the beginning of the Irish Presbyterian church, this revival was the first of its kind, at least the first recorded, to rise up in the British Isles. Converts swooned, panted, shouted, and wept. Scandalous sinners forsook their evil ways, embraced Christ as their Savior, and radically changed their lifestyles. Without doubt, one hundred years before Whitefield began to preach, Ireland experienced its own awakening. This awakening established traditions of revivalism and enthusiasm, traditions that would inform the course of Irish religious history and, through immigration, American religious history for the next four centuries.

Therefore I have begun with this Ulster revival: the early attempts to colonize Ulster and the final success experienced when the colonists were Scots; the call of Scottish ministers and the allowances made for them by the Scottish bishops; the disenchantment first experienced by the Scottish clerics. The ministers earned early rewards as revival broke out near Antrim and quickly spread to the rest of the eastern Ulster counties. Furthermore, this enthusiasm soon appeared in the west of Scotland, moving back and forth, to and from Ireland, as ministers and congregants traveled in both directions. During the early seventeenth century the Scottish church was in transition toward a reformed Protestant community. Thus the rhetoric and popular rituals that characterized this early period laid the foundation for future religious development in Ireland and Scotland. For in 1638, after all the nonconformist ministers in Ireland had been replaced by strictly Anglican priests, those same nonconformist clerics would help bring the Scottish church to rebellion, casting all the prelatists out of the Church of Scotland and returning it to the original Knoxian Presbyterianism.

By the beginning of the eighteenth century, "Knoxian Presbyterianism" would become an easy catch phrase, representing to Scots the true Scottish, Protestant tradition. This tradition would serve the Stuarts, the prelatists, and the sectarians, only to be invoked by competing clerical parties throughout the next two hundred years. In other words, the concept of Knoxian Presbyterianism assumed a mythical status, growing, no doubt, out of the legendary stature of John Knox, the first Scottish reformer, and the belief that this Presbyterianism explicitly followed Knox's dicta. A historian may be tempted to examine this seventeenth-century Protestantism in terms of its compatibility with Knox's own theology and polity. However, this sort of evaluation is far less important than the fact that the people believed certain practices and ideas were consistent with Knox's perceptions and others were not. Moreover, people made decisions based upon these beliefs. The reformed Knox did not arrive in Scotland until 1560, and his church was not established in writing until 1590. Thus, during the first three decades of the seventeenth century, one observes not the continuation but the development of a legendary tradition. All through this century, out of the conflicts between Presbyterian and Anglican, Protestant and Catholic, English, Irish, and Scottish, the tradition of Knoxian Presbyterianism was explored and developed. By 1715 a single Presbyterian tradition was perceived to have been set in stone in 1560 and to have continued through the years, unscathed and unchanged. Of course, this perception of a stable tradition was not new in 1715; it had been shared by the reforming clerics of 1638, who believed that they were returning to their origins when they were actually still establishing the nature and boundaries of those origins. It was during this period of religious instability that James I decided to invite Scottish colonists into Ulster.

The formal program for the planting of Ulster with Scots followed more than four centuries of Scottish settlement in Ulster. As early as the thirteenth century, Scottish mercenaries, initially hired by Irish chieftains to fight clan wars, were compensated with Irish land and Ulster brides. The MacDonnels, both the Highland and the Isle branches, had expanded into Antrim and Down by marrying into the Irish sept leadership. For three hundred years the two Celtic peoples grew together; the Highland Scots were quickly assimilated into the Irish culture and thoroughly accepted into the intricate social network.[2]

Until the sixteenth century, English expansion into Ireland had been in the south only and was concentrated within the pale surrounding Dublin. All through the Tudor reign, Ireland fell, county by county, under the supposedly civilizing influence of England. Yet the two primary methods used, the regranting of territory to those chieftains who surrendered to the Crown and the conquest and colonization of those areas ruled by less amiable leaders, proved unworkable in Ulster. Reinforced by alliances

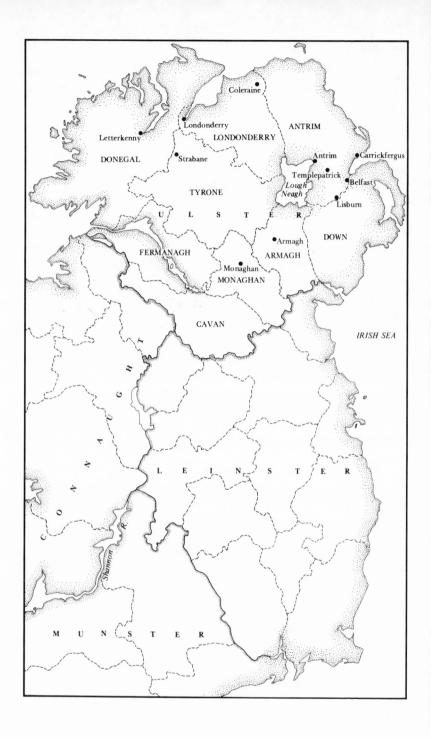

with the Highlanders, Ulster, though decimated by war, pestilence, and famine like the rest of the country, remained indomitable.

Conditions in Ulster changed dramatically during the first years of the reign of James I. Elizabeth's Irish campaigns had been particularly brutal, and some of the natives had emigrated to the continent. More important, in 1603 the chieftain in Antrim and Down, O'Neill, had found himself in serious difficulties with the English government and had traded two thirds of his land to the Scottish Lords Hamilton and Montgomery in exchange for their favorable influence at the English court in his cause. Since Hamilton and Montgomery were both good businessmen, there soon developed two flourishing colonies of Scots in these two counties. Further, the Earls of Tyrone and Tyrconnel, angered by the Anglican bishops' claim to church lands from their holdings and frightened by the king's decision to arbitrate a dispute between Tyrone and O'Cahan, fled the country. This "Flight of the Earls" in 1607, with thirty to sixty of their followers, left the remaining six counties of the province in great disorder, with large tracts of land unoccupied. Into this vacuum James sent his planters.

The greatest difference between these plans and previous English colonization efforts was the use of Scottish planters. Although the wealth of Scottish lords could not begin to compare with that of English nobles, toward the end of the sixteenth century some Scots had grown rich enough to be able to invest in a colony. The early success of the Hamiltons encouraged such investment, and James, himself a Scot, was willing to open the opportunity to his countrymen. These Scottish investors, following the example of Hamilton, invited Scottish laborers to settle their territories. Since land was cheaper and wages were higher in Ireland than in Scotland, many farmers were willing to relocate. Moreover, many of the English planters, initially following the example of Montgomery, who simply re-leased his land to the resident Irish tenantry, began to take on Scottish tenants. The Scots were excellent tenants, improving the land and raising enough grain to feed the entire provincial population.

It is always difficult to characterize the unidentifiable classes. Most historians agree that Ulster immigrants were Lowlanders, not Highlanders, mostly from the west, and mostly Protestant. Logically, migrants would be from the landless, poorer classes. Stewart's oft-quoted portrait is derogatory:

> [F]rom Scotland came many, and from England not a few, yet all of them generally from the scum of both nations, who, for debt, or breaking and fleeing from justice, or seeking shelter, came hither, hoping to be without fear of man's justice in a land where there was nothing, or but little, as yet of the fear of God.[3]

With the large number that probably migrated from the border counties, it is likely that many were fugitives; a small percentage were actually banished to Ireland. Records do show a flourishing black market in stolen goods, especially cattle, carried on across the Irish Sea between the northeastern ports of Ulster and western Scotland.[4] A more sympathetic description was given by an emigrant Scottish minister, John Livingstone. He reported his own Irish congregation as tractable but ignorant. Yet he did not arrive until 1630 and may have inherited some benefits from previous pastoral labors.[5] Robert Blair, the best known of the early Presbyterian ministers, provided perhaps the most equitable account:

> [A]nd albeit Divine Providence sent over some worthy persons for birth, education, and parts, yet the most part were such as [were driven by] either poverty, scandalous lives, or at the best, adventurous seeking of better accommodation. . . .[6]

This colonization of Ireland by England may have been motivated originally by visions of enhanced power and wealth; and by 1600 the native Irish had so violently and ruthlessly battled English domination that the colonization may have become a personal vendetta for some of the English leaders. Nevertheless, the public justification for this massive colonization effort was the civilizing of Ireland, a task that included replacing the Roman Catholic church with a Protestant organization. James I was the first British monarch to undertake seriously religious reform in Ireland. In the beginning, the Protestant church in Ireland was a peculiar hybrid reflecting the competing identities of the monarch. James had been raised by staunch Presbyterian lairds, his early years coinciding with the height of Knox's influence. After ascending the English throne, however, James, flattered by English bishops and excited by the extensive power allotted by the English church to the monarch, quickly forsook his Presbyterian upbringing and embraced the prelatist church. He also insisted upon anglicizing the Scottish church, reestablishing once again an episcopacy in Scotland. Many of the early seventeenth-century bishops in Ireland were Scots, as were most of the early Ulster Protestants, products of the Knoxian church cast in a prelatic mold.

It had taken only thirty years for the Knoxian reformation to sweep completely through the Scottish Lowlands. John Knox had returned to Scotland from Geneva in 1560, bringing with him a rigid, austere Calvinist theology and a radical Presbyterian polity. The Scottish reformers, with Knox, began to organize the Scottish church into congregations (out of the original parishes), presbyteries, synods, and a national organization to oversee the various units. In the Knoxian model, ministers, highly educated, were ordained to serve particular congregations; furthermore, ordained elders, chosen from among the most worthy congregants, would maintain spiritual and civil discipline with the support of both the ministry and the civil magistrates. The reformed Scottish church emphasized the Word of God, stripping all liturgies of what reformers called superstitious features. Gone were choirs, hymns, Latin, vestments, kneeling, statues, novenas, feast days, all to be replaced with Bible reading, catechism, long erudite sermons, and solemn psalm-singing.

Of course, the Scottish reformation did not progress as smoothly as this brief account implies. Several factions supported reform, often for political or economic reasons, and in 1560 it was not certain that the Calvinists would ultimately succeed.[7] For example, William Ferguson suggests that the decline of Catholicism in Scotland was directly tied to the decline of the power of the Guise family, the French regent for Mary Stuart.[8] Yet even if the initial victory of the Presbyterians was at risk, the events of the next three decades rendered the Scottish church Protestant. By the 1590s James VI himself supported the Presbyterian network in order to consolidate his own power, and the Presbyterian church, free of prelacy, became the Church of Scotland. James' political maneuvers, which he followed despite his preference for episcopacy, are only one indication that this Presbyterian system, set up as the established church, was becoming part of the Scottish national identity. As Ferguson concludes, "The special strength of Scottish presbyterianism lay in the fact that it embraced the idea of one national church . . ."[9]

That the concept of Presbyterianism was still uncertain or flexible can be seen in the progress of religion during the reign of James I. When in 1610 James introduced bishops into the Scottish church, he did not seriously disturb the quality of religion. Bishops were easily defined as permanent moderators of the presbyteries and synods, and therefore were able to fit right in with the established administrative structures. Moreover, no minister previously ordained by presbyters was required to undergo re-ordination by a bishop. In 1617, however, when James made his long-promised (and only) visit to Scotland after becoming King of England, he was determined to bring the Scottish church to conform more closely to the Church of England. Thus, he cajoled and/or forced the Scottish church to accept the five infamous Articles of Perth, which provided for kneeling at the reception of the sacrament, private communion, private baptism, the confirmation of children by bishops, and the celebration of the festival days of Christmas, Good Friday, Easter, Ascension Day, and Pentecost. The passage of these articles split the Scottish church. Although church histories are rather vague, if not silent, on the large number of ministers who must have reconciled themselves in quiet obedience to these acts, it is nonetheless evident that while a proportion of churchmen enforced these articles, an equally determined contingent of ministers refused to give up the symbols of reformation.

Out of this complicated, strife-ridden society came the early leaders of the Church of Ireland in Ulster. Here several ministers, ousted by the bishops of Scotland for nonconformity, were permitted by Robert Echlin and Andrew Knox, the sympathetic Scots-Irish bishops of Down and Raphoe, respectively, to exercise their ministry free from Anglican strictures. Indeed, extensive allowance was made for at least two of the pastors. When John Livingstone was to be ordained by Andrew Knox, the bishop accepted the condition that he ordain Livingstone as one among many presbyters. Further, the bishop allowed Livingstone to cross out anything objectionable in the ordination service; but Livingstone "found that it had been so marked by some others before that I needed not mark anything."[10] And Bishop Echlin's conversation with Robert Blair, as recorded by Blair, displayed a flexibility all its own. Said Echlin:

> "I hear good of you, and will impose no conditions upon you; I am old, and can teach you ceremonies, and you can teach me substance. Only I must ordain you, else neither I nor you can answer the law, nor brook the land." I told him that was contrary to my principles; to which he replied, both wittily and submissively, "Whatever you account of Episcopacy, yet I know you account a presbyter to have divine warrant; will you not receive ordination from Mr. Cunningham and the adjacent brethren, and let me come in amongst them in no other relation than as presbyter?"[11]

Blair had no objections and on the above terms accepted ordination in the Church of Ireland.

Church historians have inferred much from Robert Blair's satisfaction with the ordination arrangements. Blair was an early leader in the Scottish church, a well-known scholar, and regent at the University of Glasgow during the 1610s. He has been perceived as a man of high Presbyterian principles because he resigned his university post in opposition to prelacy and conformity to the Church of England. Thus Blair's acceptance and ordination have been used as evidence of the flexibility of the early Irish Protestant hierarchy. Yet only two bishops were really flexible at all, and even Echlin and Knox suspended and deposed nonconforming Scottish ministers after 1633.

What Blair's acceptance of ordination at the hands of the Bishop of Down (with other ministers) does indicate is the flexibility of the ministers and the transitional nature of the ministers' perceptions of the church.

Though based within the tradition of Knoxian reform, their church was still in flux, alternating between an appointed, hierarchical structure and a representative form of church government. Indeed, bishops were not finally ousted from the Scottish church until 1690. Blair's willingness to serve in the Church of Ireland demonstrates his ability to fit into an episcopal communion, provided that certain key rituals and beliefs could be retained within individual parishes. These included the invalidity of a prelate's ordination in favor of several fellow presbyters, the rejection of private administration of sacraments, the reception of communion in a sitting posture, and the dismissal of church festivals as idolatrous. While clinging to these as the symbols of Knoxian reformation, Blair was willing to accept the presence and administrative function of bishops. Pearson's description of this era as "Prescopalian" is an appropriate description and analysis.[12]

It may also have been that the bishops were influenced in their actions chiefly by their inability to acquire qualified clergy from England. It was quite unlikely that educated ministers, secure in their own parishes, would have been anxious, or even willing, to take on the onerous task of ministering to the "scum of both nations." In need of a qualified clergy, perhaps the bishops were happy to accept the nonconforming Scots.

Edward Bryce, the first Scottish pastor settled in Ulster, was installed at Broadisland in 1613. Within twelve years, the eastern counties of the province enjoyed the labors of fifteen dissenting ministers, all but two of whom had been ordained or educated in Scotland. (John Ridge and Hubbard [first name unknown] were both educated at Cambridge and ordained within the Church of England.)[13] Because these men were officially members of the Church of Ireland, they were able to minister unscathed and unobstructed, occasionally preaching Protestantism to the Roman Catholic natives as they disciplined their own, sinful people and organized them into congregations. According to his gifts, each pastor had a limited degree of success in both fields. Progress was slow, though in some cases consistent, until 1625. In that year the Church of Ireland in Ulster experienced a major transformation: the first signs of the Six-Mile-Water Revival.

The several accounts of this phenomenon that have been passed down diverge on many details, but they all agree that the instigator of the revival was James Glendinning.[14] He has been described variously as "godly but eccentric" and "all but distracted."[15] Many of his own colleagues and most seventeenth- as well as twentieth-century church historians agreed that Glendinning, though he held a master's degree from St. Leonard's College of St. Andrew's, was not among the more intellectually gifted. Robert Blair reported having gone to hear Glendinning preach at Carrickfergus, a major northern port, military outpost, and population center. Blair had been led to expect a learned scholar and was astonished to hear him "citing learned authors whom he [Glendinning] had never seen nor read." In response, Blair asked Glendinning if "he thought he did edify that people?" Blair continues, "He was quickly convinced, and told me he had a vicarage in the country, to which he would retire himself quickly."[16] Thus did James Glendinning choose to retreat to Oldstone and preach to the country surrounding Antrim.

Most recorders of the history of this period have expressed consternation and amazement that a man so unworthy, so apparently unintellectual and potentially unstable, could have had the kind of success that Glendinning experienced. The answer lay in the fact that unlearned though he was, the man was a powerful preacher, "having a great voice and vehement delivery."[17] John Livingstone reported that Glendinning had changed his method of preaching on the advice of Robert Blair, though not as Blair had advised. Glendinning "fell upon a thundering way of preaching and exceedingly terrified his hearers."[18] Inspired by the "great lewdness and ungodly sinfulness of the people," Glendinning "preached to them nothing but law, wrath, and the terrors of God for sin," and, as Stewart so neatly concluded, "in very deed for this only was he fitted, for hardly could he preach any other thing. . . ."[19] In appropriate response to such hellfire attacks, his hearers began to faint, cry, and scream in the terror of their consciences. They began to ask the question that so many ministers longed to answer: What must we do to be saved?

Unhappily, Glendinning seemed unable to answer that question. While fully conversant with the horrors of sin and punishment, he lacked, according to Blair, the full understanding of the gospel that would enable him to settle his listeners and "satisfy their objections."[20] His parishioners continued to writhe, but no one experienced the joy of conversion. Fortunately for the revival, the neighboring ministers came to his aid. John Ridge, scarcely a mile away at Antrim, suggested the formation of a monthly lecture to be set up at Antrim. The preachers included Ridge, Blair, Robert Cunningham of Holywood, and James Hamilton of Ballywalter; Glendinning, recognized as a problem, was excluded. The hearers came from all directions. These meetings began with a sermon on Thursday evening, continued with sermons all day Friday—three in the winter months and four during the summer. The ministers used this time for general consultation among themselves concerning the administrative as well as the spiritual affairs of their congregations. Although Irish church historians described these meetings as the forerunner to presbyteries, their greater importance lay in the constant attendance of hundreds of people gathering to hear the gospel preached and to pray for twenty-four hours at a stretch.

The swift progress of the revival that spread through the Ulster Scottish congregations was recorded by Blair, Livingstone, and Stewart. By 1630 the excitement had traveled beyond Antrim and Down. Three- to five-day prayer meetings involving thousands of people became a frequent occurrence. And, as with any successful revival, problems of counterfeit conversion and misunderstanding began to develop. Since Blair was one of the first involved in these events, he seemed to have been more preoccupied and troubled by the accompanying problems. He reported a false awakening at Locklearn, where "persons fell a-mourning," "afflicted with pants like convulsions." However, the afflicted displayed no "sense of their sinfulness" nor "any panting after a Saviour," so it was concluded by all to be a false imitation. Robert Blair seemed to have played a role common to Jonathan Edwards and Gilbert Tennent a hundred years later in keeping the revival honest. When sent to preach among the people of Locklearn,

> one of my charge, in the midst of public worship, being a dull and ignorant person, made a noise and stretching of her body. Incontinent I was assistent [*sic*] to rebuke that lying spirit that disturbed the worship of God, charging the same in the name and authority of Jesus Christ not to disturb that congregation; and through God's mercy, we met no more of that work, the person above mentioned still a dull and stupid sot.[21]

The ministers had to be careful that they did not mistake physical agony and enthusiastic reactions for true spiritual awakening. As witnesses, Blair, Livingstone, and Stewart were all careful to note the radical change in the lifestyles of the converts, for they had to defend against accusations "made against us, as if we had taught the necessity of a new birth by bodily pangs and throes."[22]

Not only were these instruments of revival combating the hostile opinion of hierarchy, they were torn themselves by this peculiar form of spiritual renewal. They had desperately sought success among their congregations and were now asked to accept it with conditions bordering upon chaos. Livingstone recounted stories of enthusiasm at Broadisland, responding to all accusations with the simple statement that Edward Bryce was a "godly, aged minister." Livingstone further admitted that some persons did pant and such during services, but that most of the ministers, especially Bryce, discountenanced it, since people "were alike affected whatever purpose was preached."[23] Yet when discussing Andrew Brown, an old deaf-mute who had been notoriously vicious and loose, Livingstone spoke favorably of the same type of emotionalism that he was wont to discount—Brown wept at sermons. Of course Livingstone carefully explained that Brown had left his old ways and companions, joined with religious people, attended church services, and went to prayer alone. Nevertheless, his emotional response was seen as proof that he had experienced a true conversion and enlightenment.[24]

Livingstone had longed for revival. His own report of his first months in Killinchy (1630–31) hinted at an envious despair over everyone else's revival. When Livingstone first arrived he discovered an established Presbyterian-style congregation with three elders doing their best to maintain discipline among a "cold" and "indifferent" people. Livingstone wanted revival, and some insight into the methods by which such response was elicited can be gleaned from the account of Walter Stuart's conversion. Stuart, a "heinous profligate," claimed that he had been awakened by the preaching of Livingstone. Meeting his pastor one day while on a solitary walk in the woods, Stuart was quoted as saying

> I ha'e been deaf to reproof—I ha'e been blind to my misery. . . . I ha'e loved sin—I ha'e laughed at mercy in my mad moments—I ha'e drunk deep the cup o' pleasure—o' pleasure—o' misery —I ha'e drunk it greedily—I thacht it sweet—I called it pleasant—I ha'e drunk it to madness—it is now the venom of asps, gall and wormwood are in the cup. I went, sir, to hear you preach; your words awoke within the state of lost souls—of their endless misery—of their hopeless regret—of their eternal woe—a fearfu' dread has settled on my soul—I seek rest, but I can find nane.

Livingstone responded with inspiration; Walter Stuart at last found peace in conversion, and the revival began in Killinchy.[25]

The revival spread not only through Ulster's eastern settlements, but beyond Ireland to western Scotland. During the same years, David Dickson, pastor at Irvine, was attracting great numbers into the church. "Multitudes were convinced and converted."[26] People from communities several miles distant from Irvine were attending Dickson's sermons and joining in his communion services. During 1630 "Stewarton sickness" became a catch phrase describing the phenomenon. Stewarton and Shotts, small towns near Irvine on the western coast of Scotland, incidentally among the closest points to Ireland, were the centers of massive religious events. Blair remembered the "daft people of Stewarton," reporting that "numbers of them were at first under great terrors and deep exercises of conscience, and thereafter attained to sweet peace and strong consolation."[27] For many of the recorders of these events, the zenith of the revival seemed to have occurred not in Ireland but in Scotland, following a communion service at Shotts.

> The communions were the chief seasons of preaching among the Presbyteri-
> ans, and they were used as opportunities for collecting together adherents of
> the party, not only ministers, but professors also, as the lay people were
> styled.[28]

One contemporary observer reported carefully a series of events that has
become familiar to students of revival. The date was 21 June 1630, the
Monday after a communion service.

> [T]here was so convincing an appearance of God, and down-pouring of the
> *Spirit,* even in an *extraordinary way,* that did follow the ordinances,
> especially the sermon, on *Monday, June 21,* with a *strange unusual motion
> on the hearers,* who, in a great multitude there were convened of divers
> ranks, that it was known . . . near 500 had at that time a discernible *change*
> wrought on them, of whom most proved lively christians afterward . . . and
> truly this was the more remarkable that one after so much reluctance, by a
> special and unexpected providence, was called to preach *that sermon* on
> *Monday,* which then was not usually practised, and that *night before* by most
> of the christians there was spent in prayer, so that *Monday's* work, as a
> convincing *return of prayer,* might be discern'd.[29]

This period was quickly established in Scottish tradition as a time of
great religious success. When the Great Awakening came to Scotland in
1740, several participants—notably James Robe, the primary apologist for
the eighteenth-century phenomenon—looked back upon these years as a
legitimization of the contemporary outbreak. Even the Associate Presbytery, Seceders from the Church of Scotland who claimed to follow the
true Scottish reformation tradition and who opposed Robe and his followers, could discount the eighteenth-century revival only by establishing
its essential difference from the seventeenth-century event. In their *Act,
Declaration and Testimony* the Seceders affirmed the activities in western
Scotland during the 1620s and 1630s as the beginning of the return of
strong Presbyterianism to the Church of Scotland, a precursor to the ecclesiastical reform of 1638.[30]

More important, the accounts themselves began a tradition of narrative that emphasized the providence of God, the spontaneity and inspiration of the preachers, and the immediacy of the effects upon the hearers.
Blair told many stories of his extemporaneous preaching, sermons he felt
compelled to deliver. He recalled one instance in which so many people
had attended an event that they could not fit into the church building. They
would therefore have been left without reward for their coming unless Blair
had stepped into the breach. At other times, both he and Livingstone were
plagued with long days and nights during which they could not prepare a
sermon for the coming Sabbath; suddenly, either at midnight Saturday or

fifteen minutes before the service began on Sunday morning, they would
be inspired. Moreover, these sermons had the greatest effect upon the
greatest number of people, as during the services at Shotts, when five
hundred people were converted at once, proving that God was taking care
to pour out the divine spirit.

Where did the revival actually begin? It is now almost impossible to
tell whether Ireland or western Scotland evidenced the first signs of transformation. Without doubt, James Glendinning was awakening his hearers
as early as 1625, and this was perceived as a new and disturbing phenomenon; Blair said that very few months passed before John Ridge established
the Antrim meeting in response to the outbreaks. Clearly, too, the identifiable occurrences in Scotland at Shotts and Stewarton occurred in 1630.
Nevertheless, the precise year of the beginning of the revival remains
unknown. Certainly Blair's report of a notable providence, a good harvest
that followed a day of fasting held in response to too much rain, was given
as an explanation for an upsurge in spiritual activity among his congregation. In fact, Blair was hard pressed to convince one such follower who

> took up an erroneous opinion, that there was need of no other means to be
> used b[ut] prayer, whatever ailed soul or body, young or old, corn or cattle
> [that] it was a tempting of God to neglect other means.[31]

This occurred in 1624. Further, during 1624, eight communion services
were celebrated by Blair and Robert Cunningham together which were
attended by all members of both congregations, and this year also experienced an increase in the use and efficacy of congregational discipline.[32]
Although the emotional, enthusiastic outbursts were described as beginning in 1625 under the preaching of Glendinning, some Presbyterian ministers were succeeding before this. While there are no similar reports from
western Scotland for the period preceding the revival, it is possible that
success followed those ministers opposed to the anglicization of the Scottish church.

The key to this question involves the difference between a revival of
religion, as the contemporary clerics called the phenomenon of that period,
and revivalism. Seventeenth-century ministers used the term revival to
designate a revitalization of piety among the people of Ulster and western
Scotland. Part of this general revival, indeed the major part, was reflected
within those rituals identified by the term *revivalism*, by which I mean
rituals focused upon conversion and characterized by a highly charged
emotional and physical, supposedly spontaneous, response to deliberate,
organized efforts to stimulate that response. The revitalization of piety
began sometime during the 1620s in northern Ireland and western Scotland,
but the 1625 reports of Glendinning's activities are the earliest indications

of revivalism as such. What is most important to understand is that for eight years during the first half of the seventeenth century, revivalism, in all its emotional fervor and emphasis upon conversion, flourished in both Scotland and Ireland, nourished by the ministers as the means of returning their people to God.

During this period of intense religiosity, piety increasingly revolved around the sacraments. The interpretations made by the early Protestant reformers of Roman Catholic liturgical services, especially the Mass, always emphasized the superstitious or magical aspects of the rituals. Yet the Protestant clergy's attempts to turn the people's minds away from the superstitions of popery should not be perceived as a rejection of sacramental theology. On the contrary, the Scottish reformers actually wanted to return the sacraments of baptism and, especially, the Lord's Supper to a ritual centrality that they no longer enjoyed. Baptism had become a private ceremony, celebrated only among family and friends; the Protestants held that it should be a community event, performed privately only in cases of extreme necessity, for example, the imminent death of an infant. Even worse from the Protestant perspective, communion had come to be celebrated irregularly, if at all, in parishes. Scarcity of priests, scarcity among priests of conscientious practitioners, and an emphasis upon the unworthiness of the laity had removed participation in this sacrament from the common experience of most of the laity. The demand of the Scottish church that ministers celebrate the Lord's Supper with their parishes at least once per year ensured frequency; its recommendation that the sacrament be celebrated as often as four times annually was radical.[33]

Consider, too, the intensity of the discussion surrounding ritual symbols and practices. Robert Blair walked out of a communion service during his tenure at Glasgow because the celebrant insisted that recipients kneel. Four of the five Articles of Perth dealt with sacramental theology in some form, for example, by allowing private participation, by calling for an unacceptable ritual such as kneeling, or by affirming confirmation as a sacrament, which the followers of Knox argued was unbiblical. Throughout the sixteenth and seventeenth centuries, sacramental theology became a subject of hot debate and a primary method of establishing one's religious identity, not only as distinct from Roman Catholic, but from English Anglican (or, depending upon one's affiliation, from Scottish Presbyterian) as well. While some points of difference concerned accusations of superstition or debates involving biblical justifications, others were matters of community identification. Scottish reformers saw the sacramental celebration as an affirmation of religious community; thus the reformers strenuously objected to private occasions.

This focus upon the communion service as a major religious-com-

munity event grew dramatically during the revival. The communion service became the primary vehicle for spreading excitement. In the various
accounts of this period, the Lord's Supper consistently surfaced as a central
event in the lives of ministers and people. Not only did people react
violently to the various parts of the communion experience, but the ritual
itself became a tradition embodying many of those qualities that would
characterize revival meetings for the next two hundred years.[34]

Communion services were very large. They were scheduled so that
the celebration in one parish would not conflict with that in any other.
Neighboring congregations were then notified of the coming celebration of
the Lord's Supper and invited to attend. Outsiders were encouraged to
attend, not only by the central focus which their own church placed upon
the event, but by the concrete fact that their minister would probably be
asked to assist the minister in whose congregation the sacrament was being
held. For example, in 1624, Blair and Cunningham each celebrated four
services in their own congregations, each assisted the other, and both
congregations attended the eight services. Noting another such instance,
Blair counted at one service in 1630 "as many people outside the church as
in"; he estimated that the building held four hundred families at least.[35]
John Livingstone wrote that he held services only twice yearly, explaining
that it was not necessary to do it more often, for nine or ten parishes within
twenty miles of each other all held communions at different times, with the
neighboring ministers assisting at the various meetings.[36]

In Scotland the size of the participation was even greater than in
Ireland. At a service held in Stranraer in 1638, several ministers counted
five hundred families coming from Ireland (their own ministers having
been deposed) in addition to the hundreds of persons resident along
the west coast.[37] Throughout the seventeenth century large services became the standard. At least one presbytery in Scotland carefully assigned
ministers to assist one another in their communion services—as many as
eleven clergymen in one service. When, in 1697, the Church of Scotland
attempted to reduce the number of communicants at a single service "for
the preventing prophanation of the Lord's day at Communions by great
confluences of people," the Synod of Glasgow and Ayr required each
parish to hold a service at least once a year, restricted pastoral assistance to
three or four ministers, and forbade any other ministers to attend.[38] Yet it
remained impossible for the institution to control what had clearly become
a popular ritual. Sessions records for congregations in both Scotland and
Ireland included acquisitions lists preparatory to a general celebration of
the sacrament. They would include such items as four dozen loaves (or
perhaps a bushel of flour), eight to eleven gallons of wine, a gallon of

brandy, four cups, four flagons, and so forth. The elders clearly expected to serve large numbers of people, and the size of the collections made during those services indicated that they were rarely, if ever, disappointed.[39]

In addition to being large, communion services were long. They began on Saturday, with an entire day devoted to sermons of preparation. Sunday was wholly devoted to the distribution of communion itself. This would usually begin at mid-morning with an hour-long sermon and an exhortation/invitation. The distribution would often last until past dusk, sometimes making the service ten to twelve hours long. The elements were distributed not from an altar, but around "a table, ten yards long, where they sit and receive the sacrament together like good fellows."[40] During the entire day the ministers took turns distributing, preaching during the distribution, and preaching to the crowds left to wait outside the building until room could be made for them at the table. Monday was devoted entirely to thanksgiving. Although only the home congregation could be guaranteed to participate in all three days of the service, many individuals from neighboring congregations would arrive early, having left their homes late Friday evening or early Saturday morning, and remain through Monday, leaving only in the evening after the service was completed.

Furthermore, the rituals were actually structured to prolong this period of intense community piety for ten days or more. Preparation in a congregation might begin as early as the Sunday before, with a special preparatory sermon delivered by the preacher. Catechising would dominate the work of the preacher and the lay elders for the rest of the week. One day of the preceding week, usually a Wednesday but sometimes a Thursday, would be set aside as a fast day for preparation. In many congregations the hours of thanksgiving went beyond Monday, continuing well into Tuesday. Naturally, there would be another fast day set aside for thanksgiving. Finally, the next Sunday was frequently celebrated as a postcommunion Sunday, with special sermons and collections. The question arises as to how people could devote as much as two weeks to this activity, neglecting their farms. It can only be noted that communion services were rarely held in mid-spring or in the autumn, and were held most frequently in late May through early July, right after planting, and in November, right after harvest. Occasional references in the sessions records do indicate that this was a conscious decision.

During the early period of Irish protestant history, other factors worked to intensify and prolong the religious experience of the laity and the clergy. The monthly meetings begun at Antrim by John Ridge in an effort to balance the horrifying effects of Glendinning's preaching were

frequently followed by a communion service either at Antrim or at one of
the neighboring congregations. Following a congregational fast day held
either Wednesday or Thursday, hundreds of people would arrive in Antrim
Thursday afternoon to partake of the twenty-four hours' worth of sermons
delivered at the meeting. The preparation for the Lord's Supper would then
begin either Friday evening or Saturday morning, and so on.[41]

Further, the paucity of trained ministers and the threat of deposition
that hung over the nonconforming clergy encouraged the people to take full
advantage of the services they did have. Thus Livingstone could report that
people would

> with great hunger wait on the ordinances. I have known them that have come
> several miles from their own houses to communions to the Saturday sermon,
> and spent the whole Saturday night in several companies, sometimes a
> minister being with them, sometimes themselves alone, in conference and
> prayer, and waited on the public ordinances the whole Sabbath, and spent
> the Sabbath night likewise, and yet at Monday's sermon were not troubled
> with sleepiness, and so have not slept till they went home.[42]

The same passage also records the development of yet another aberration
traceable directly to the dearth of ministers: the assertiveness—the praised
assertiveness—of the laity. Not only did lay leaders direct the secular
affairs of the church and advise the pastor concerning spiritual matters;
they actually led prayer services. Stories like that of Hugh Campbell began
to be told. Ripe substance for the generic conversion narrative, Campbell
had been known as a great sinner: he had fled Scotland after committing
murder. However, God had "caught" him and transformed him into an
"eminent and exemplary Christian." Stewart went on to say that following
his conversion, Campbell

> became very refreshful to others who had less learning and judgment than
> himself; and, therefore, invited some of his honest neighbours who fought
> the same fight of faith to meet him at his house on the last Friday of the
> month, where, and when beginning with a few, they spent their time
> in prayer, mutual edification and conference of what they found within
> them. . . .[43]

These traditions of all-night prayer meetings and meetings conducted by
inspired lay men and women would continue to influence and intrude upon
the workings of the Presbyterian churches in both Ulster and Scotland for
years to come. Among substantial numbers of the clergy, this lay leadership engendered a general fear of losing control to the laity and an open
mistrust of those religious types who threatened to claim sacred authority
for themselves. As mentioned previously, the end of the century saw

efforts, albeit unsuccessful, to place these services under a stricter institutional, clerical control.

The communion service itself followed a fairly well-established pattern. Since common homiletic practice was to preach according to a regular schedule of Old or New Testament texts, verse by verse, chapter by chapter, until an entire book of the Bible was completed, a presacrament sermon ought to have been quickly recognized by the people, because its text would undoubtedly digress from the schedule. Beginning with sin, the pastor would preach upon the horrors of sin and the terrifying punishments awaiting the hardened sinner. These sermons might have begun consciously as early as the Sunday before, or they may not have begun until the preceding Saturday. Underlining this preparation were the efforts of the elders to eradicate all scandal from the congregation. At Killinchy, John Livingstone began the practice of holding a special meeting of the sessions Saturday morning in order to identify any scandalous persons before the actual Sunday celebration. These persons were then summoned before the elders and told to appear before the congregation that very afternoon and acknowledge their sins. All who did so were publicly absolved; all who

> would not come before us or, coming, could not be induced to acknowledge their fault before the congregation, upon the Saturday preceding the communion, their names, scandals, and impenitency, were read out before the congregation and they debarred from communion; which proved such a terror that we found very few of that sort.[44]

In those congregations where such public cleansing directly preceded the Sunday communion, the individual acts of penance followed by absolution served as a symbolic reconciliation for the entire community. Even those congregations that did not experience the public repentance of sinners as a necessary part of their three-day service practiced the distribution of the tokens that had to be presented for admission to communion. On the fast day preceding the communion, tokens were distributed to all communicants in good standing. The token represented an open declaration from the community attesting to a person's general fitness to receive. In the case of refusal, each church member present observed the careful alienation of the unfit. The unfit included not only fornicators, adulterers, and perpetrators of violent crimes, but also drunkards, cantankerous neighbors, slanderers, liars, and profaners of the Sabbath. Moreover, all those who were given tokens were reminded through the preparatory sermons that while they appeared worthy, in fact they probably were not, and only they and God could really know the state of their own souls. Just as the community was publicly purified so that as a whole it would not profane the

Lord's Supper, so individuals were instructed to examine carefully their own consciences and prepare their souls to receive the wonderful, purifying gift of the sacramental presence.

Sermons preached during the communion itself emphasized both the atoning acts of Jesus Christ and the possibility of salvation available to each of the faithful. The emphasis changed from "prepare yourselves" to "open your hearts." On these days, too, were heard the most graphic depictions of the life and passion of Christ. The sacrament was, after all, a memorial of Christ's sacrifice, which the people were encouraged to ponder with compassion and pity. In one sermon Christ was described as "our blessed and eternal Saviour stuck under the 5th ribe by one of the soldiers," and the hearers were instructed to "think much upon his death think you are at Golgatha at the foot of his Cross waloing in his blood. . . ."[45] Yet, despite such verbal illustrations, the emphasis was not so much upon guilt and gore as upon the joy of the salvation accomplished through Christ's death and resurrection.

Monday's hearers were treated to two equally important themes: thanksgiving for both the gift of saving grace and the celebration of the Lord's Supper, and warnings against relapse, accompanied by exhortations to desert the world and follow Christ. Thus, over the three-to-ten-day period, communicants were called to consider their sins and the consequent punishment; to recall the saving act of Christ's atonement and open their hearts to grace; to experience personally the joy of salvation; and to leave the ways of the world and follow Christ, that is, to sanctify their lives, guarding against falling into the same old, sinful ways.

Of course, the Scots-Irish did not invent this structure. Students of early Protestantism recognize it as grounded in the Augustinian concept of conversion that so many Calvinists embraced. It followed precisely the paradigm of conversion that the English Puritans so avidly sought, with one essential difference. In the case of the Puritans, in both old and New England, conversion was primarily the experience of the individual believer. Many communities, especially in New England, would have no such need since they were composed totally of the visibly converted.[46] Among the Scots and the Irish Presbyterians, however, the possibility of a visible-saint church was deliberately rejected. Thus, all but the infamously wicked were considered members. Since each parish would have unsanctified members, the community itself was impure. From this standpoint the community must experience conversion, and within and by the community each individual would be nourished. Members of both churches experienced conversions during the early seventeenth century; but in the case of the English, these experiences were intensely personal, while in the case of the Scots-Irish, they were vibrantly communal.

This period of heightened religiosity soon ended in Ireland. As early as 1632, reports of enthusiasm had been passed along to the English church authorities, probably by bishops in Scotland envious of the ministers' success. The first threats of deposition were warded off primarily by James Ussher, archbishop of Armagh from 1625 to 1656 and sympathetic to the Presbyterians' work. Due to Ussher's intervention only the most prominent Presbyterian clergy were attacked. In 1632 they were suspended for six months. Most of the ministers acquiesced to the penalty while Robert Blair, mustering political influence from all possible sources, personally placed their case before Charles I. Defending the ministers solely against the charge of enthusiasm, Blair won the argument, and he and his colleagues were restored to their ministries.[47]

This, however, was but a brief respite in the long turmoil that lay ahead. The year 1633 saw the rise of William Laud to the See of Canterbury, this appointment renewing and strengthening his determination and efforts to force conformity within the Anglican communion throughout the British Isles. Within that same year, Thomas Wentworth, later earl of Strafford, was appointed Lord Deputy over Ireland. Wentworth, absolutely loyal to Charles and the English government, ruled with the intent of destroying, in Ireland, every force—political, economic, and ecclesiastical—that challenged the ultimate authority of the English crown. A competent, though unscrupulous, administrator, Wentworth did effect the centralization of power within the English governing institutions in Dublin. Yet his machinations and manipulations, frequently resulting in the betrayal of the very interest groups he pretended to champion, earned Wentworth, over his seven-year tenure, the mistrust of an amazingly diverse collection of Irish residents. He undermined the economic ascendancy of the Anglo-Irish merchants, many of them London-based, by chipping away at their trade monopolies. He alienated the native Irish elite by promising an extension of Charles' "graces"—religious tolerances for the Roman Catholics—in return for cooperation, and then he rewarded that cooperation with the restriction and final abolition of those graces. Most relevant to this story, he drove the Scottish clergy and many of their disciples from Ulster with his carefully constructed policy to anglicize, and thus stabilize, the prelatic Church of Ireland.

At the Convocation of 1634, Wentworth, under the direction of Laud, maneuvered the adoption of the Church of England's Thirty-Nine Articles to replace the Irish Articles. The 104 items that constituted the Irish Articles had been drawn up and passed in 1615 through the influence of James Ussher, who at that time was professor of divinity at Trinity College, Dublin. The articles promulgated a Calvinist theology and inclined toward a Presbyterian form of church government; in fact, they provided a key document in the formulations of the Westminster Assembly of the 1640s.[48] Although most of the clergy did submit to the new articles in 1634, a Calvinist tendency within the Church of Ireland was indicated by the Convocation's political decision to leave the Irish Articles unrepealed.

Wentworth was faithful to the interests of the Anglican church, reinforcing episcopal authority and abetting the filling of sacred coffers, his policy necessarily turned against the Presbyterian Scots. From a Presbyterian perspective, the history of the 1630s was little more than a series of acts of religious persecution culminating in the passage of the "Black Oath" in 1639. Blair, George Dunbar, and Josias Welsh, the cleric who continued the revival at Oldstone following Glendinning's mad departure in 1628, were finally deposed in 1634; Bryce, Cunningham, Livingstone, Ridge, James Hamilton, and Samuel Row were deposed in 1636. Furthermore, those ministers refusing to subscribe the Thirty-Nine Articles and the new Irish Caroline Canons, also passed in 1634, were subject to criminal prosecution. One example cited by John Barkley illustrates the inflexibility and fierceness of this persecution. Already deposed in 1636, Robert Cunningham, of Holywood, fled to Irvine where he died in 1637. Following his death, the High Commission summoned Cunningham to appear before them. When he did not appear, he was fined twenty pounds, despite the Commission's notification of his death. The court then seized his estate as security for his appearance, with no consideration for his widow.[49]

Flights such as Cunningham's were not limited to the clergy. In 1636 a large group of Ulster Protestants, upon invitation from Massachusetts Bay, outfitted a ship, the *Eaglewing*, and set sail for New England. The voyagers reached Newfoundland waters before storms and bad winds forced the ship's return to Ireland. Dispirited and frightened, many passengers followed the return migration to Scotland. Here the Presbyterian clergy, increasingly able to confront the episcopal hierarchy, conducted the very services that the prelatic church had forbidden in 1617 when the nonconformists fled to northern Ireland. It was during these years that the five hundred or more were said to have crossed the Irish Sea in order to participate in a communion service at Stranraer, for their own worship had been curtailed as their ministers were deposed and driven away.

The fact that Presbyterian worship services were being held openly in Scotland during this period should have warned the bishops that their authority was no longer respected, their power no longer feared. The numerous restrictions placed on the Church of Scotland in an effort to force conformity and bring about ecclesiastical unity with the Church of England alienated a burgeoning contingent among the moderate, previously docile clergy and inflamed a significant proportion of an irascible, culturally

independent Scottish laity. Unconcerned by challenges to church polity
and theological integrity, these Scots raised their hackles at the thought of
anglicization. They were certainly unhappy with the persecution of their
countrymen in Ireland, and when Laud tried to require the use of the Book
of Common Prayer in Scotland, the Scottish church rebelled. Led by a
radical party that included the Irish exiles Blair and Livingstone, the clergy
assembled at Glasgow in 1638. With one sweeping movement, the passage
of the National Covenant, the Glasgow Assembly ousted the bishops,
removed English innovations from their worship, and restored Presbyterianism, conforming as closely as possible to Knox's 1590 codes. Several
bishops were not only deposed but excommunicated for such faults as
celebrating festival days and hanging crucifixes in their rooms. Throughout
the Scottish Lowlands, congregations openly signed this covenant in a
ritualized renunciation of England and the Anglican church.[50] The official
Irish response was quick and decisive. Wentworth effected the passage, in
1639, of the Black Oath. All men and women over sixteen years of age
were required to swear that they would never oppose the king and that they
would renounce the National Covenant. Removing any doubt as to his
intentions, Wentworth exempted all Roman Catholics from the oath while
exacting a parish-by-parish list of all Presbyterians so that none would be
passed over. Of course some did oblige, but the insidiously vague and
unlimited nature of the oath drove many Scots to refuse. The outrageous
fines, as high as five thousand pounds per person, and the long imprison-
ments exacted as punishment brought entire communities to leave their
homes and flee to Scotland.

Presbyterian historians have judged as fortuitous the harsh measures
that led the Scots to return to Scotland. Although Wentworth fell from
power in 1640, his policies had so aroused and angered the native population that the Irish broke into armed rebellion in 1641. At first directed
against the Anglicans/English and restrained to lootings and captures of
towns, the rebellion of 1641 grew to massacre proportions, sparing no
Protestant, regardless of age or sex, and subjecting its victims to cruel and
barbarous tortures and deaths. While modern scholars continue to dispute
fact versus fiction concerning the countless atrocity stories, enough has
been proven to establish, through the size, speed, and violence of the
rebellion, the intense anger and hatred that had been nursed among the
native Irish.[51]

In 1642, in response to the Irish rebellion, the now Puritan English
parliament sent a Scottish military force to settle the country. With the
army came five Presbyterian chaplains who found themselves among a
very few Protestant ministers in Ulster. The five organized their troops into

congregations with sessions, and themselves into the first official Irish presbytery. News quickly spread throughout the province of Ulster, so that the tiny presbytery was inundated with requests for preaching and sacraments. The presbytery, in turn, petitioned the General Assembly of the Church of Scotland for assistance, and some of their clergy, including Blair and Livingstone, were sent on three-month itinerant tours of the region. The Irish Presbyterian church was now firmly established, based, like the new Church of Scotland, upon the Scots Confession, the Second Book of Discipline, and the Book of Common Order—the foundations of the creed, discipline, and liturgy of the original Knoxian reformed church. Congregations were reestablished, and elders were ordained and empowered to maintain discipline in the absence of clergy. Scandalous persons not under a pastor's care could be received by the elders upon public repentance, although lay authority was limited in that public repentance was not required of sinners unless they were under a pastor's care or unless they wanted to participate in the sacraments. The presbytery further encouraged those conforming ministers who had survived the rebellion to join them. Many ministers did, surprisingly, join the newcomers even though the presbytery demanded evidence of each minister's conversion in the form of public penance among the laity for the erroneous practice of conforming. Finally, and of great symbolic import, many individuals who had sworn the Black Oath acquiesced to the demand for a public acknowledgment of that pledge as a crime.[52]

The Scottish church was feeling its way, trying to retain old traditions while coping with the challenges posed by political instability. Faced with a possible alliance between Charles and the Scots, the English parliament was pressured into signing the Solemn League and Covenant, a political document affirming the rule of Parliament and establishing the Presbyterian institutional system throughout Britain. Thus English divines, in their efforts to create a new confession of faith, found themselves impelled to accept Scottish direction. In this manner the decade produced two sets of documents around which the Scots, at home and in Ireland, could unite. The Westminster Confession, the Longer and Shorter Catechisms, and the Directory were fully acceptable replacements for the 1560 prescriptions, and subscription to these treatises would soon become a prerequisite for church membership. The Solemn League and Covenant, devised as a pact between Puritan England and Scotland, became an immediate symbol of Scottish national identity, an identity that included Presbyterianism.

Men and women throughout Scotland and Ireland participated in public ceremonies, often held on the Sabbath, during which each individual signed the document. The covenant held such import for the Scots that after Charles I had signed it, the Scots refused to wage battle against him. These Presbyterians were horrified at his execution, not merely as the murder of a king, but as the murder of a covenanting monarch. When Charles II was restored to the English throne in 1660, the Scots agreed to subject themselves to his rule because he had signed the Solemn League. Yet, in order to reestablish episcopacy in Scotland and Ireland, and in direct opposition to the promises he had made, Charles II demanded in 1662 that the Solemn League and Covenant be publicly burned and renounced. And in 1733, a full hundred years later, when the extremist evangelicals seceded from the national church, they would justify their secession with the assertion that they, and not the official Church of Scotland, had remained true to their foundations laid out in the National Covenant of 1638 and the Solemn League and Covenant of 1643.[53]

The church that met Cromwell and his sectarian army was fiercely independent and nationalist. As in every other area under siege, in the Scottish Lowlands and in northern Ireland the diverse responses to civil war presented points all along a continuum from loyalty to the Stuarts, through devotion to the Solemn League, to enthusiasm for the sectarians. Throughout the late 1640s and 1650s Cromwell would try to gain the support of the Scots with a combination of force, in the form of restriction and persecution of the opposition, and reward, in the form of government stipends for supportive clerics. Yet neither Cromwell nor Charles ever really understood that, for the Scots, faithfulness to God meant faithfulness to Scotland, which, in turn, meant faithfulness to the reformed Scottish church as designed by John Knox. The Scots refused to be English. Because neither Charles I, Oliver Cromwell, nor Charles II could understand this nationalism, each failed to conquer Scotland.

United as they were against the encroachment of Englishness, the Scots were divided among themselves. Robert Baillie recorded the serious conflicts that disturbed the functioning of the Church of Scotland during this period, with ministers battling over the respectability of accepting the salaries, the clergy, and the organization granted and/or imposed by Cromwell's government. The issue was not whether it was acceptable to give in to the English administration; everyone assumed that this was not acceptable. The real point seems to have been whether the acceptance of stipends from Cromwell and attendance at those councils were a betrayal of Scottish interests to England. The conflict so overwhelmed the interest and energy of the majority of the clergy that the warnings of leaders like Baillie, that such battles would only weaken unnecessarily an institution that had hitherto enjoyed a basic unity, went unheeded.[54] The Scots may have been undefeatable, but they were also incapable of winning.

During this period of tribulation other intrachurch differences were repressed, yet the differences did surface now and again. When the Irish exiles returned to Scotland, they settled throughout the Lowlands, east and west, bringing with them the popular rituals that they had enjoyed in Ireland. They were most welcome in the west, and most of the banished ministers were settled in congregations along the western coast. Robert Blair was delighted with his new charge at Ayr, as were the people with him. So delighted were all, in fact, that when Blair's academic reputation led to a call to St. Andrew's, city of the university, Blair refused to go. In good establishment fashion the General Assembly of 1638 ordered Blair to accept the call to the collegiate parish, "yet they [Ayr] have kept still Mr. Blair, almost by force." The transfer never did occur.[55]

The reality of divisions among Presbyterians concerning religiosity and spiritual expression was underlined by a peculiar dispute aired before the General Assembly at Aberdeen in 1640. The Laird of Leckie, a recent exile from Ireland, was accused of praying in such a manner as reflected badly upon Henry Guthrie and other ministers "who did not affect their wayes." Such prayer was apparently delivered at prayer meetings organized and led entirely by lay persons, without the formal consent of minister or church. Guthrie and the magistrates tried to suppress these meetings, without success, for the exiles held that these meetings were simple gatherings for family prayer. Both sides indulged in a goodly amount of railing during the assembly. The church wished to prevent any serious division, and thus hoped to cast no reflections upon the pious of Ireland. An ad hoc committee decided that the partisans generally agreed on the essentials and that reconciliation was possible. Henry Guthrie would preach in favor of family worship; Blair, Livingstone, and John Makclellan would speak against night meetings and other abuses complained of. According to Guthrie, however, Blair did not decry the meetings with adequate vigor; Guthrie therefore refused to fulfill his part of the agreement.

As the battle continued in the assembly, the phrase "Irish innovations" kept appearing, as if such practices had been observed only in Ulster, when western Scotland had probably experienced a fair share. However, Guthrie was a minister in Stirling, central Scotland, and his allies came from the surrounding area and from Edinburgh, where the leadership of lay persons at prayers and spontaneity in worship were rarely seen, let alone encouraged. Differences in ritual practices were finally reduced to accusations of "unconverted minister" and "prideful layman." At one point Samuel Rutherford produced a syllogism: "What Scripture does warrant, ane Assemblie may not discharge; bot privie meetings for exercises of religion Scripture warrants," that earned the response "Lord

Seaford would not have Mr. Samuell to trouble us with his logick syllogisms." Still another committee tried to resolve the dispute and managed to overcome the mistrust of all involved with the following conclusions: First, all agreed that read prayers, as opposed to spontaneous, were not unlawful. Second, all agreed that no one was permitted to expound Scripture except ordained ministers or student/probationers who had been approved by their presbyteries. Everyone also had to agree that families were permitted to worship together. The problem came in the desire of Guthrie's side to state explicitly that family worship included only members of the immediate family and other members of the household. Of course no one could refute the definition as such, but Baillie did point out that this would not resolve the problem of the private meetings, since the disruptive meetings were not family gatherings. Yet Guthrie maintained that he would be satisfied with this public statement, and the truce was drawn.[56]

Uneasy as the truce was, both sides were prepared for battle during the 1641 assembly. Despite Andrew Ramsey's sermon of reconciliation from Psalm 122, David Dickson appeared to many "too passionately vindicating the credit of religious people from unjust slanders, and urgeing repentance of such Ministers, who, with their conformitie, had brought lately our Church to the brink of ruine."[57] Though some may have found this an unfair attack upon Guthrie and his friends, Guthrie's opponents clearly believed that their innovations were rooted in the Scottish tradition and that the easterners were guilty of the sin of anglicization. Small aspects of ritual, such as "omitting Glory to the Father, Kneeling in the pulpit, discountenancing read prayers . . ." that so troubled the ministers of Edinburgh as innovation were, for the exiles of Ulster and the western Scots, part of their Knoxian heritage.[58] Once again the Edinburgh ministers were brought to suppress their aversion to the "privie meetings" and to accept the innovations, and Guthrie and Leckie, with their mutual accusations of slander, were brought to reconciliation. Soon political tribulations would force the squabblers to abandon their battle for more serious challenges to their power.[59]

Thus, during the critical years between 1638 and 1643, the first invocations of a barely established Knoxian tradition were seen. The reformers of 1638 self-consciously returned to a known set of documents, insisting upon a complete return to sixteenth-century practices. They perceived the Stuart monarchy as destructive of true religion and therefore hoped to avoid every innovation or adaptation that had occurred during the previous forty years. Yet they would find that this was not possible, nor would it have been desirable for these particular reformers, since many of the developments of the seventeenth century strengthened the religious identity of the Scottish and Scots-Irish people.

With the new importance granted by the people to the communion service, the primary sacrament assumed a centrality in the community, and with this came a recognition of the importance of baptism and marriage (performed by an acceptable minister). Thus a new authority was granted to, or perhaps assumed by, the church and its clergy due to their power over these rituals. In addition, the community through these rituals had established a process of community conversion or purification, a process that served, at this point, to enable individuals to clarify their own religious experiences and turn toward God. At the same time, the focus remained upon the community, not the individual, underlining the importance of the church hierarchy, institutional organization, and clerical authority within the Knoxian system.

Yet the impact of this era upon Presbyterianism goes beyond an increase of power and authority accorded to the established church, back to the new centrality of piety itself. Simply put, the people were interested in attending religious services, and the services they wanted to attend were revivalist. Such rituals involved massive numbers of participants collected indoors and out; they lasted for several days, sometimes as long as two weeks. Individuals responded openly and emotionally to provocative preaching and prayers, seeking the ultimate goal of conversion. Lay participation was encouraged during these services, as well as in the general spiritual life of the parish, and this was a power that the laity were loath to relinquish.

When the General Assembly was petitioned in 1642 by the new presbytery at Carrickfergus, the assembly considered carefully the options as it defined them, both in terms of personnel and tenure. Baillie reported that "It was the Assemblie's care, to beware lest all the men who went over to that land should be in danger, in the first settling of that Church, to favour any differences from our Church."[60] His summary reflected a peculiar perception of the differences existing within the church. First, Baillie did not realize that the differences were already too firmly rooted to be destroyed by a lack of sympathy. Second, he did not see that these differences, these popular rituals foreign to the Edinburgh community, were shared with the western Lowlands—a truly Scots-Irish phenomenon. Finally, he assumed that "our Church," the national Church of Scotland, was an Edinburgh-centered church. Not only was this assumption a hopeful underestimation of the strength of the deviants, it would soon be overturned. In less than thirty years these deviants would become nationally recognized as the true relict of the Scottish reformed church.

2

Scottish Conventicles and Irish Communions

The seventeenth century was a period of proverbial chaos that swept the British Isles. The Scots had hoped that the Solemn League and Covenant, signed by both the English parliament and the Scottish assembly, would serve as a contract of mutual aid and alliance. They also expected that once Charles I had been convinced to sign the covenant, all would settle down in unified peace. Unfortunately, the sectarian party in England had ousted the more moderate Presbyterians, and the sectarians had little interest in negotiating an amicable settlement. So Charles was beheaded, the sectarians renounced the covenant, the Scots were alienated, and the civil war continued another ten years.

In the face of this great political stress and betrayal, the Presbyterian churches of Scotland and Ireland struggled. By 1650 neither church had been established in a firm, Presbyterian base. The first official Irish presbytery was founded only in 1642, and that among military chaplains serving the Scottish regiments. Before that initial institutionalization, a Presbyterian-oriented people had been served by Presbyterian-oriented clergy within the confines of the Anglican Church of Ireland, and that accommodation was prevalent for about ten years only. Mid-century Scotland was in a Presbyterian period that cyclically alternated with episcopal ones; the present Presbyterian reformation had begun only twelve years before. For all this seeming instability, Knoxian Presbyterianism had been nourished in some form in Scotland or Ireland since 1560; and the Presbyterians were developing a keen sense of what this tradition was, as well as a determination to maintain that tradition intact.

The maintenance of this tradition through the latter half of the seventeenth century was a fascinating struggle. Challenged from without by English political demands, from within by conforming and extremist clerics and lay persons, those Presbyterians who would in the end be recognized as the standard bearers developed a variety of survival strategies. A few of these strategies were developed during the Commonwealth period, during which Presbyterianism as such was organized in Ireland and the Westminster Assembly documents were established as the new credal standards of the Church of Scotland. As the restoration of Charles II to the British throne brought renewed persecution to the Presbyterians, specific rituals were employed by the nonconforming ministers to defy the civil authority and reinforce the Scottish tradition. In order to understand the full range of this cultural response to the Stuart monarch, it is necessary to examine the Scottish and Irish communities separately, for the English government applied very different levels and kinds of pressures to each communion, resulting in two different survival strategies and thus two different churches.

The continuance of Presbyterianism in the face of alternating non-recognition and oppressive attention involved the commitment of ministers and people to ideas, institutions, and traditions. Whatever might be continued in the religious experience of the people could then resurface as the church after the tribulation had ended. The means by which ideas and traditions were continued was partly determined by what was acceptable in the past; but the Scottish Protestant past was rather brief and lacking in variety. During the latter half of the seventeenth century, the community faced obstacles and pressures not seen before. The few traditional responses available were not adequate, and the community was forced toward innovation. Thus the church developed new rituals in direct response to new challenges; the new quickly became old, and other practices were tried, while some of the old ones dropped away and others became established as the Presbyterian way. What could be observed during this period of intense political and religious stress was that those traditions that fed the emotions also encouraged community and provided an active identity in opposition to the perceived enemy. Past rituals that have been described as revivalist survived and combined with new activities, also revivalist in character, to continue the fight of Scottish and Scots-Irish Presbyterianism. By 1690 these evangelical tendencies all but overwhelmed the other forces within radical Presbyterianism. Moreover, while the voice of moderation was soon raised again, there would always be a fierce, self-conscious core of radicals (or reactionaries?) who would equate evangelical revivalism with Knoxian Presbyterianism, with Scottishness, with ultimate truth.

An important segment of the developing evangelical tradition was the introduction of the Westminster Confession into the Scottish church as the most perfectly written expression of the Christian faith. In signing the

Solemn League and Covenant in 1642, the new parliament in England had committed itself to defend all who entered into the covenant, to protect the life of the king, to eradicate popery and episcopacy, and to restructure the Church of England "according to the word of God, and the best Reformed Churches."[1] To this end Parliament called an assembly of divines to meet at Westminster in July 1643. All the participating ministers, except two French Huguenots, were English and had been ordained into the Anglican church: thirty-one were doctors of divinity; thirty-nine were masters of arts; and at least twenty were tutors or fellows at Cambridge or Oxford.[2] To augment this collection of English intellectual talent, the Scottish assembly was, in August, invited to send commissioners to attend and advise the Westminster Assembly. These Scottish commissioners were not granted any voting privileges; and while six were appointed, only four took any active part in the deliberations: Alexander Henderson, Samuel Rutherford, George Gillespie, and the young Robert Baillie.

The interaction of the English and Scottish ministers was tarnished by their awareness of Parliament's political need to retain Scotland's support and their knowledge that the Scots were attached to their church as an essential piece of their identity, a non-negotiable condition of the Solemn League. For the Scots, the "best Reformed Churches" included Swiss, French, and Scottish Calvinists, certainly not Lutherans or Anglicans. Additionally, the English Calvinists, that group broadly identified as Puritan, had never experienced the opportunity, or the necessity, to organize a national church. The only other example that the Westminster Assembly could have followed was the new colony in New England, but beyond the simple question of whether the English were even knowledgeable about Massachusetts Bay, it remained true that Massachusetts could provide very little concrete assistance in the political cause. Scotland's resources were extensive and its support was essential. Inexperience combined with expediency must have upset the balance in favor of the "powerless" Scots. The members of the assembly took an oath to "maintain nothing in point of doctrine, but what I believe to be most agreeable to the Word of God; nor in point of discipline but what may make most for God's glory, and the peace and good of this Church," and yet most of their decisions were favorable to the Scottish position.[3]

The assembly sat four years, during which they produced a Confession of Faith, Larger and Shorter Catechisms, a Directory for the Public Worship of God, a Form of Government, and a Directory for Ordination. The very Scottishness of these products was attested by the quick adoption of these writings as replacements for the formulations of Knox that had always been used. Although the acts of adoption in both Scotland and

ROSS

Moray Firth

NAIRN

ELGIN

BANFF

ABERDEEN

INVERNESS

KINCARDINE

GRAMPION MOUNTAINS

FORFAR

PERTH

NORTH SEA

Perth •

St. Andrews •

ARGYLL

Auchterarder •

CLACK-
MANNAN

FIFE

KINROSS

Stirling •
STIRLING

*Firth of
Forth*

Kilsyth •

DUMBARTON

LINLITH
GOW

Edinburgh •

HADDINGTON

Paisley •
Glasgow •

Shotts •

EDINBURGH

RENFREW

Clyde R.

BERWICK

Cambuslang •

Hamilton •

Carluke •

BUTE

Stewarton •

Lanark •

LANARK

PEEBLES

*Arran
I.*

*Firth of
Clyde*

Irvine •

SELKIRK

Ayr •

AYR

ROXBURGH

DUMFRIES

Dumfries •

KIRKCUDBRIGHT

ENGLAND

WIGTOWN

Stranraer •

*Solway
Firth*

*North
Channel*

Ireland carefully tied the Westminster documents to the 1560 forms, the writings of the Westminster Assembly quickly became the standards borne by the Scottish and Irish Presbyterian churches. All people were to be educated through the Westminster catechisms, polity was structured according to the new Directory, and all ministers and elders would be required to subscribe the Westminster Confession as the confession of their faith. On the other side, after the restoration of Charles II in 1660 the English Puritan organization, reduced to dissenter status, never accepted the creeds or disciplinary codes of the Westminster Assembly. As early as 1680 these documents were seen by English nonconformists as imperfect, restrictive, and not English, while for Scotland these documents became a symbolic remnant of the true reformation, a statement of orthodoxy that, within twenty years, would become traditional and unalterable, more solidly linked with the Presbyterian tradition than the 1560 codes that had been replaced.

While the Scots were constructing an agreement with the English parliament through the machinations at Westminster, the Presbyterians in Ireland were building a church network. When in 1642 the people heard that a presbytery had been erected, applications from various districts throughout the north of Ireland arrived demanding supply of preaching, catechizing, and celebration of sacraments. Congregations with active sessions were organized in fifteen towns in the area surrounding Carrickfergus in that first year alone. As the knowledge spread, the number of requests increased. Petitions were sent to the General Assembly of Scotland for ministers, some specifically requesting that once-resident clergy, such as John Livingstone and Robert Blair, be returned to their Irish congregations. However, with a shortage of clerics in Scotland as well, the best that the assembly would provide was a sequence of missionaries who did three-month preaching/sacrament/discipline tours of the countryside.

As episcopacy was prohibited and Presbyterianism was the form of the national church of England, Cromwell, during the 1640s, was friendly toward the Presbyterian clergy. As soon as the Scottish church delivered its public statement condemning the execution of the king, however, Cromwell's religious policy was remodeled to protect his political and military interests. In 1650, an Engagement Oath, calling for a renunciation of Charles I and the succession rights of the Stuart line and for allegiance to the Commonwealth, was required of all clergy. Those refusing to take the oath, and most Irish Presbyterians refused, were driven from their pulpits; many returned to Scotland for safety. A 1653 act had banished them from the country, but such an act could not be enforced. When Henry Cromwell became Lord Deputy of Ireland in 1654, he ended the persecution of the ministers and granted stipends of a hundred pounds per year to any minister who applied, including many Presbyterians and former Anglicans. The five years of Henry Cromwell's rule allowed the Presbyterians to regain their strength. In 1654 the community was divided into the three Presbyteries of Antrim, Down, and Route. The Presbytery of Laggan grew out of Route in 1657, and in 1659 Tyrone was formed out of Down. These met in synod, or great presbytery at Antrim, as circumstances required.

What is astonishing is that these years do appear to have been, for the Ulster Presbyterians, a period of stable growth. Settlements organized themselves into congregations that selected elders and petitioned the presbytery for ministerial supply. The records of the Antrim Meeting read like those of a presbytery during a time of civil tranquility. Most of the time and energy of the Antrim Meeting was devoted to the discipline of congregations, the oversight and evaluation of ministerial candidates, and the supervision of pastoral relations. Because the civil government provided annual stipends for the clergy, congregational disputes were far less violent than they would become fifty years later. The major concern of the meeting was that the minister be provided with a house and glebe and that the congregation promise him emotional, spiritual, and additional financial support as a token of the other two. Perhaps the scarcity of the clergy rendered people more supportive and less critical. Certainly the presbytery was less demanding, having developed no system of examination for licensing preachers. Instead, probationers were permitted to preach among the large number of vacant congregations; they underwent trials only as a preparation for ordination to a specific charge.[4]

This is not to say that the final trials were not rigorous. Each candidate for ordination, during the course of his trials, would deliver several homiletic exercises upon several biblical texts, then move on to the presentations of a learned sermon and a popular sermon, also on assigned texts. After these exercises had all been approved, the candidate would prepare a Latin exegesis and an English commonhead, that is, a Latin and an English exposition upon questions posed by the presbytery. Sometimes the two would be on the same question of theology or church polity, but generally the two topics would be different. Finally, each candidate answered spontaneously questions on cases of conscience, church chronology, and the reconciliation of contradictory places of the scriptures; and each demonstrated a working knowledge of Greek and Hebrew. Nevertheless, at this point a candidate was required to pass trials only once—as a qualification for ordination—whereas Presbyterian probationers in Scotland (and in Ireland after 1660) usually had to pass two series of trials—a preliminary examination for licensing, followed by another for ordination.

The scarcity of clergy, directly following upon an earlier period of total dearth, helped to create an atmosphere of commitment among lay congregants. When the first presbytery of Antrim was established in 1642, congregations reorganized themselves and requested supply. The presbytery then aided the people in developing an institutional structure that could survive with intermittent attention from itinerant missionaries who felt no commitment to any one community. The elders of the churches, selected from among the laity according to various criteria that reflected social and moral respectability, became the general overseers of piety and conduct. With complete authority over the churches' civil affairs, the elders supervised the collection and distribution of the poor monies, as well as the upkeep of the church property. They further enjoyed the preliminary responsibility for discipline, holding hearings and prescribing penalties and fines without the supervision of a pastor. Throughout the year the sessions, or elders' court, would ask the presbytery to send a clergyman to preach, celebrate the Lord's Supper, baptize the children, or assist them in resolving a particulary difficult breach of morality.

Traditionally, the lay leadership had always played an active role in congregational affairs. Their authority over financial and property decisions at the congregational level had stood in contradistinction to the pastor's authority over spiritual matters. Naturally, the elders were often perceived as spiritual leaders by and among the laity and the official church. Yet within the Presbyterian structure, the elders' spiritual actions, even as an official, elected sessions court, were performed under the direct supervision of the pastor. Without a settled minister, the elders' authority, and the laity's perception of that authority, was greatly enhanced. The clergy during these years were less important to the daily life of the congregations. Ministers provided occasional technical expertise in disciplinary cases and lent their sacramental presence in support of the elders' authority.

The Antrim Meeting became a court of appeal to which elders could refer extreme cases. During the four years recorded in the earliest batch of Antrim minutes, 1654–58, the ministers heard cases of slander, scolding, breach of promise, and drunkenness, to cite a few examples. The small number of cases heard in these categories (slander was the highest, with six cases) indicated the success that individual sessions were having within their communities. In the parish of Templepatrick alone, between 1652 and 1654, the elders successfully handled eleven cases of Sabbath-breaking and fourteen trials for drunkenness. The offenders were accused and persuaded to perform public penance for their sins, after which they received public absolution.

The most frequent breaches of discipline were sexual. In Temple-patrick twenty-nine instances of fornication were processed, along with six cases of adultery, while the Antrim Meeting tried only thirty-nine cases of fornication, compared with forty-nine adultery cases. Since fornication was considered a far less heinous crime than adultery most congregations dealt with these cases themselves, sending only a disputed case of fornication to the general meeting, while most adulterers would be automatically referred to presbytery and synod. Of course, sexual crimes always comprised the bulk of any schedule of moral discipline because the consequences were concrete, visible, and socially disruptive. However, it was not quite so usual that sins unrelated to sex should be so thoroughly represented. Swearing, scolding, accidental bloodshed, Sabbath-breaking, drunkenness, and slander all were found in these two sets of minutes. It is within the minutes of the Templepatrick sessions that reference was made to provision of a stool of repentance, a wooden fixture placed in front of and facing the congregation, upon which the condemned sinner would sit until publicly absolved. So too, requisition lists for communion materials indicate a continuation of these large services. This was a time of commitment to a pure community, to maintenance of the good already present, and to real reform of the wicked.[5]

On 5 November 1657 the Antrim Meeting called a "Day of Fasting and Prayer" to be observed toward the mitigation of sin. The list of nine causes distributed with the call for the fast day read much like any other list of causes. "Abuse of former fasts," "uncleaness, Drunkenes, pryde, malice, covetousnes, deceipt, and swearing," "Gross Sabbath breaking," and disobedience of children and servants were among the sins recorded. Nowhere was there a hint of civil stress or instability. Instead the people were criticized for "Unprofitableness & unfruitfulness of the most part notwithstanding of the continued enjoyment of the lords ordinances, & variety of speakeing dispensations of mercie and judgement yt we have mett with." People prayed that God would continue merciful in his dealings with northern Ireland, an open statement that ministers worked, people worshiped, and Presbyterianism continued without serious external interference. Yet all was not peaceful, for in the midst of all those causes of the fast could be found hints of internal strife. Buried within the nine reasons for the fast was the prayer that God would purge the land of "hyrlings and men unsent of [who] go to preach the gospel, which hath been and may prove a snare to the people through their inadvertence and simplicity."[6] While these itinerants may have been sectarians, it is likely that they were renegades within the Scottish Presbyterian fold.

During the interregnum two competitive clerical organizations divided the Scottish church. This division, so violent and unrelenting, especially in the west, brought the church in Ireland, in the summer of 1654, to pass an act in support of Presbyterian unity. Henceforth, the Irish Presbyterian church required all candidates to disregard the differences festering within the Church of Scotland.[7] Many Scots were unable ever to accept the support of the Commonwealth government, and, being Scots, they denied that any who did accept that government's support could possibly have been acting in good conscience. Thus, three clerical groups competed for Scottish congregations: the Independents, or sectarians; the Remonstrators, a group of Presbyterian ministers so called because they remonstrated against trusting the new king without evidence of his commitment to the reformation and the covenant; and the old Presbyterians, who condemned the Commonwealth government as founded upon regicide and were prepared to trust Charles II on the basis of his signature to the covenant. Needless to say, the Independents and the Remonstrators enjoyed stipends and kirks, "but whom the Presbyterie, and well-near the whole congregation, calls and admitts, he must preach in the fields, or in a barne, without stipend."[8] Many of the presbyteries and synods throughout Scotland split, with the Remonstrators now maintaining the discipline of the established church while their opponents continued the tradition of the "true" Church of Scotland.

Despite the regular interference of Cromwell's military forces, resulting in the forced dissolution of several general assemblies and synods, disestablished Presbyterians managed to retain some of their popular authority; but this church could not survive unchanged. Originally the church of the civil state, the Presbyterian church now found itself opposed to the government. Used to preaching in churches and receiving tithes, the ministers were forced into barns and fields, left financially to the good will and support of individuals and families legally required to pay for a minister they did not want or trust. Their sacraments were no longer public celebrations but services hidden from public notice and advertised only by word of mouth. No wonder that during these years there developed a new breed of preacher—one whose style was adapted to the peculiar circumstances of his work. So Robert Baillie described one such preacher, new to Glasgow:

> He has the new guyse of preaching . . . contemning the ordinarie way of expounding and divinding a text, of raising doctrines and uses; bot runs out in a discourse on some commonhead, in a high, romancing, unscriptural style, tickling the ear for the present, and moving the affections in some, but leaving, as he confesses, little or nought to the memorie and understanding.[9]

Considering that the ranting sectarians had earned a reputation from this sort of preaching and that the Remonstrators, many resident in the west, employed such tactics, it was inevitable that the techniques earlier used by James Glendinning would infiltrate the mainstream Presbyterian community.

Not all areas of Scotland experienced active conflict. Many presbyteries in central and eastern Scotland continued to support and discipline congregations. The Synod of Glasgow was split, with each side claiming the truth for its own; but the Presbytery of Dumfries, for example, was as eager to avoid disputes as was northern Ireland. Dumfries worked throughout the 1650s to remove the sectarian influence and replace uneducated, unappointed ministers with proper Presbyterian preachers. Yet within Dumfries, this careful, methodical process provided a counterpoint to the emotional style starting to predominate over the quiet, learned method of the Edinburgh-centered community. In a contemporary description of a communion service at Kirkpatrick in Durham, it was reported that Gabriel Semple

> always employed the most lively Ministers. . . . But none was more Countenanced than himself, especially at the breaking up of the Action before the Tables in laying their Sins before them, & calling them to Humiliation for the Same; And then in praying, Confessing Sin, & engadging anew the Lord to be his, & to walk in his way; At which there used to be a great Consternation & Elevation of Spirit in the Congregation.

Semple's communions were held twice a year, and his services were attended by people from twenty to thirty miles away. "He had as much of an Apostolick Gift of Preaching, & in a thundering way, as I know of any of the Ministry."[10]

It was as if the church of the 1650s were in training for a long, difficult period of endurance after the Restoration. When Charles II was restored to the monarchy, he proved the fears of the Remonstrators reasonable when he refuted the Solemn League and Covenant, ordered the document publicly burned, and curtailed the activities of the judicatures. Appointing bishops throughout Scotland, Charles laid upon the shoulders of his new, ambitious episcopate the burden of conformity. As early as October 1660 James Sharp, archbishop of St. Andrews, wrote hopefully to the earl of Lauderdale that "the Remonstrators (excepting some few phanaticks) are hoely-silenced, the profain sentences of Discipline are reduced to a dred [of] it," and that most ministers spoke well of the king.[11]

The bishops and king were soon disappointed. From another correspondent Lauderdale heard that by 1664 most of the ministers in Edinburgh had been deposed for nonconformity. The Presbytery of Paisley in 1664 suspended most of its ministers because they had refused to attend presbytery meetings. In 1662 in Glasgow an act had been passed requiring all pastors to conform or demit their charges, and Gabriel Semple was not alone when he quietly gave up his pastorate.[12]

It is significant that Charles did not attempt to negotiate unofficial toleration for the Presbyterians; certainly they did not want it. Having twice enjoyed the status of national church and believing itself the true heir of the Scottish reformation of 1560, the Presbyterian organization had no intention of accepting the position of dissenting minority. Instead, in good Scottish fashion the Presbyterians were determined to endure the prelatic period, standing and growing in opposition to the unpopular episcopacy, adding a new dimension of militance to Presbyterian church services.

The suspended, deposed, and demitting pastors did not remain "hoely-silenced." The lessons of the interregnum stood them well. When the church edifice was placed out of reach, they could use houses, barns, even open fields, and so Semple continued his career. He took up residence with a gentleman of Corsock and delivered sermons in that house. The increasing size of his audience required the moving of his meetings to open fields. Leaving his unofficial congregation to the care of John Welsh, Semple toured the western Lowlands, preaching and baptizing the children of parents who refused to worship with men who had deserted the covenant. Semple noted that as persecution increased, adherence to the presbytery was strengthened; and when the meetings were indulged, people continued to attend despite some disagreement on the righteousness of enjoying official toleration. Semple's was a period of massive preaching-meetings and tremendous celebrations of the sacraments; his was a time when individuals "were brought to the Lord, yt had not the least Profession of Religion before, & continued in the Same."[13]

Many Scottish ministers did conform, but did this mean that they sympathized with the bishops? The Presbytery of Lanark is now known to have been the home of many field conventicles during the 1670s, yet not one was reported to the presbytery. The Presbytery of Paisley heard reports of a great conventicle at Kilmacolmb in 1666, but in the end the group concluded that the presbytery had no way to legally establish the truth of such rumors, and the investigation ceased. In July of 1674 the bishop over Auchterarder asked that the presbytery decide a difficult question of patronage; the presbytery returned the problem unsolved with an explanatory note saying, essentially, not our jurisdiction.[14] Many of the presbyteries simply adopted a policy of passive inaction: they let the bishops visit their congregations, paid lip service to the church, and held services in the face of large numbers of congregants' refusing to attend or, if they attended, refusing to partake of baptism and communion.

It seems logical that such lay behavior would be resented, and yet these moderate conformists appeared to hold no grudge. For example, during the early months of 1682 the parish churches were increasingly weakened by defections to the conventicles; discipline and order were constantly disturbed by the antics in the fields. In November the bishop over Auchterarder had sent a letter to the presbytery's moderator

> appointing him to show them that againe he must putt them in minde of what the law requires of them concerning disorderly persons in their bounds & that to the Lords of his Majesties Councill ane account must be given of their names once this Moneth & that those who are short of their duty will come to greater trouble than they are aware of. . . .

In response, John Drummond said that he knew of no one who had even heard a conventicle preacher. John Philips knew of none. Alexander Ireland could name three, and the moderator could name only one. Lawrence Mercer reported that he knew of none, as did three other clerics, and they sought a delay so that they might check with their sessions. These reports, and those of the absent members, did not provide the useful lists hoped for by the government, and these men served within a presbytery filled with conventicles.[15]

This lack of zeal on the presbyteries' part made clerical discipline difficult for a bishop to enforce. William Spence, pastor at Glendovan, presented a petition to the Presbytery of Auchterarder in 1678 asserting that the church was "labouring under sundry grosse corruptions and a horrid defection from its first purity & received constitut'ne." Although his six points in many ways resembled the standard complaints of various protestant communities at various times and places, taken together they formed a complete indictment of the current establishment. Despite the condemnation of such doctrines by the synods, Spence found that Pelagian, Arminian, latitudinarian, and popish errors "wtout any notice or censure are professed, preached & propagated to our great shame & reproach." He attacked the liturgy as adulterated "wt the mixture of humane inventions," asking the synod to prevent "privat assayes to introduce into or impose upon this Church ane unwarrantable liturgie of unsound & Useless forms. . . ." He complained of laxity in congregational discipline and of "exotick," unscriptural obtrusions upon church government. He raged against the "spirit of prophanenesse, Atheism & ungodlinesse" throughout the nation, doubtlessly calling to mind the common association of profligacy and extravagance with Anglicanism. Finally, and perhaps most unforgivably, he recommended that means be taken to heal the divisions between the prelatists and the Presbyterians; initially the established church ought to grant the other's desire with the expectation that the prelatists' "doubtful & lesse warrantable practices may be remedied for the time, & prevented for ye future. . . ."[16]

The bishop orchestrated the official response to Spence's paper, calling for conferences with Spence in order to satisfy the petitioner's scruples. By February 1679 Spence conscientiously scrupled to attend presbytery. He was described as

> Too self-willed in adhering to, as he was rash and precipitant in giving that paper. Whereby it was too Like that all this affaire has been contrived, and is yet transacted upon designe to carry on some point of privat interest against the peace and unitie of this Church.

How the members of presbytery really judged their colleague remains unknown due to the interference of the bishop, who "desired the advice of the Brethren if they did not think that Mr. William Spence deserved to be suspended." Naturally, the judicature affirmed their agreement with the bishop by suspending him; but they also, three months later, affirmed their sympathy with Spence by publicly denying that they had preached any of the errors with which they had been accused.[17]

On 14 September 1679 Spence appeared once again, this time released on bond from prison. The bishop, with the presbytery, declared his willingness to hear charges and accusations from Spence; Spence replied that he was willing to go to whatever jail the bishop appointed. Spence argued that after he had presented his paper and had agreed

> to condescend upon particular persons who had professed, preached, &c. Pelagian and Arminian errors and to adduce witnesses . . . [they passed] the sentence of deposition against him immediately and so rendring me uncapable to do anything but suffer; and therefor I hold myself no more oblidged to [as]sist in that process. . . . He was not there to answer such questions, but to offer himself to prison.[18]

That was the last presbytery meeting at which Spence appeared. He was to be formally excommunicated the following May for "a most scandalous & railing Paper," for presenting and refusing to retract that paper "containing gross reflections upon the Government both Civill & Ecclesiastick," and for perjury in writing against the "oath of supremacy & Canonical obedience both taken and subscribed." Most interesting, however, was the charge that he continued to minister and preach in Glendovan although deposed with bond given. Having refused to attend four separate meetings, Spence had left the bishop no opportunity to pass the sentence of excommunication upon him; finally the presbytery reported that such sentence had been effected.[19]

William Spence, though somewhat more vocal, was only one of many Scottish clerics who ruined the sleep and digestion of the bishops. In their long correspondence with the Duke of Lauderdale, James Sharp, archbishop of St. Andrews, and Alexander Burnet, archbishop of Glasgow and later of St. Andrews, continually complained of the exercises and successes of the conventiclers. For the first two years of his tenure Sharp had noted a general acquiescence to the new prelacy. His first alarm came in autumn 1662 when he discovered that

> some ministers in the West, Lothian and Fife have so by combination
> refused to take presentiments according to the act of parliament, or to keep
> any meetings or exercise discipline because of the lye of the Covenant which
> they hold sufficient to superseid their duty and obedience to the Lawes. . . .[20]

Sharp was surprised by their defiance and recommended immediate action. Burnet was of a more tranquil temperament. When he reported in 1664 that upwards of thirty conventicles flourished in Edinburgh, he added that he was not upset that some have

> suffered these discontents and disorders [to] swell to so great a height
> because as it will discover them and lay open their designes to the view of
> the world so (if seasonably supprest) it will give our friends an opportunity to
> express how usefull and significant their presence and service may be to the
> publick.[21]

Yet he would soon be brought to give up his hope in the world's agreement and approval of his party's policy as he discovered the depth and strength of the popular discontent.

Scarcely a month after writing that hopeful missive, Burnet found that George Buchanan's *De jure regni apud Scotos,* a history challenging the legitimacy of the Stuart line, had been distributed throughout western Scotland. Six or seven of the most virulent opponents of the prelatic church, including John Livingstone, had fled to Holland, from which vantage they could publish such disruptive books and serve as martyrs for the Scottish cause. Although the people continued to be distressed about the unequal favors granted to various nobles such as Lord Argyle, the bishop of Glasgow had hopes of avoiding persecution of the malcontents and keeping the peace.

> Most of our vacancies are filled, many of our seditious ministers are laid
> aside, and even the most obstinate of our remonstrating Lairds begin now to
> buckle; and yett I am sure very little or no severity hath been used.

As a resident of Glasgow the bishop should have known that the west would never pray quietly. Eight months after his report of successful peacemaking, Burnet noted that Welsh, a fugitive, Cruikshanks, allegedly the recent translator of Buchanan, and several other preachers had attracted

> great multitudes of people together to their seditious meetings and baptized at some places twenty at others more of the children of such as live within the parishes of orderly and obedient ministers.

The bishop complained that neither the sheriffs nor any other officers, excepting two lords, one time each, had "so much as offered to interrupt them or to seize or apprehend the ringleaders that seduce & deceive the well-meaning multitude. . . ."[22]

With the outbreak and growth of these conventicles, the government tried various approaches and solutions in unsuccessful attempts to curtail them. When the magistrates instigated legal proceedings against conventicle followers, the trials influenced more people to join them and to "regard it that they are under the conduct of an high Commissioner." As fugitives from the Commission, "those that keepe the hills and glens in Galloway doe much infest the Stewarton of Kirkindbright and borders of Nithsdale, and have dealt so severely with our ministers that most of them are forced to flee. . . ."[23] Sharp was most concerned at a report that Michael Bruce was keeping a conventicle within six miles of St. Andrews, in the neighborhood of a conforming minister. Persons of all classes had been described to him as disapproving of "those extravagances," but they were afraid, or simply unwilling, to interfere. He found "those disaffected people generally more and more to embolden themselves to doe and say what their humor leads to," and that "in civil matters we find the preservation of impunity to be the great muse of disorder."[24]

As the 1670s progressed, the conventicles grew larger, more violent, and more public—in some cases the rebel ministers actually preached from the church pulpits. Sharp was rather misguided to call "pitiful" and "insignificant" those men who were so successfully accomplishing "a work to runne this country into confusion."[25] At this point several nobles advised the king to try toleration, as persecution was only binding the conventiclers more firmly together. For all these hopes, Burnet reported that where in Glasgow only seven or eight had been suspected of keeping field meetings, there were now three times that many. "It is hard to determine whether lenity or severity have been most sleighted and despised by them; for both seem to heighten their pride and prompt them to new and greater insolences."[26] The ministers conducting these meetings had almost established full parish operations. In addition to preaching and baptizing, these ministers published banns and married couples and kept sessions with elders who would advertise the ever-changing place and time of the meetings. By 1679 any conventicle preacher could acquire a license to function, and by 1679 no field preacher bothered to obtain one.[27]

Not content with civil disobedience, some of these communities came to be armed. As early as 1676, Burnet complained that his own ministers and their followers were assaulted, beaten, and plundered.[28] The few years of indulgence quieted matters to the extent that people continued for six years or so to be content with their conventicle activities. However, when James, Charles's brother, began to influence the king's policy during the early 1680s, and certainly after he came to the throne in 1685, he attempted to strengthen the power of the episcopacy and grant some toleration to the Roman Catholics. The Scottish countryside broke into armed rebellion.

Colonel James Douglas, conducting one of the military campaigns in western Scotland for the king, received in 1685 a series of letters and orders that conveyed the breadth and variety of the trouble caused and the successes enjoyed by the covenanting rebels. In January Douglas was instructed to discover the murderer of the minister of Carsfarne and give protection to the regular clergy. Later in the month he was appointed by the Privy Council to go to the area where rebels had been taken prisoner and "do justice" so that they might be an example to all. In March there was report of rebels in Dumfries, stealing weapons from people's homes; but they had remained in town until eleven the next morning, and the people in the surrounding areas did not rise in defense. The entire population was judged disaffected from the government, and Douglas was given carte blanche to seek out the rebels and to judge them as he pleased.

In Ayr rebels were foraging the countryside openly; Douglas was appointed, along with six others, to punish the commons who saw and did not report the rebels as instructed by the council's proclamation. He was further commanded to take bond for the appearance of the heritors before Parliament. Enforcing an oath of abjuration and quelling religious discontent, Douglas was provided with English Jacobite and Highland troops and an unlimited authority to control by whatever means and force necessary.

> You are to search persons and apprehend all Rebells fugitives resetters disloyal persons and others who disturb the peace and quiet of his Maties government, And to cause Immediately shott such of them to death as you find actually in Arms.[29]

The strength and power of Douglas's and other troops notwithstanding, the rebels, with the apparent support of the people, proved able to endure, if not dominate, long enough to help the English overthrow of the Roman Catholic James in favor of the Protestants William and Mary. Since the

Scottish Anglicans had taken an oath never to bear arms against James, they not only ended up opposed to the new king, but were subjected to the humiliation of seeing those very rebels, the conventiclers, returned to full participation and leadership in the establishment Church of Scotland.

What was this rebel communion that not only survived but flourished during those twenty-eight years of Stuart reign? Clearly it was identified by its membership as the Scottish communion in opposition to the anglicized church of the prelates. The accusations of William Spence told part of the tale. The harshness of Calvinism in opposition to the soft hints of Pelagius and Arminius; scriptural Presbyterian government against the unbiblical bishoprics; the plain goodness of Protestant preaching and not the frippery of the Anglican liturgy; commitment to an austere lifestyle rather than expensive subjection to fashion—all these features differentiated the two churches, according to the Presbyterians. Yet, for all these rational justifications, the patterns and tenors of opposing traditions lay deep within.

For the conventiclers, the new prelacy had its roots in the old prelacy established by James I, an establishment that reeked of anglicization and betrayal, a weak edifice that in Scotland had once been easily overturned. The faith of the rebels, on the other hand, was self-consciously traced directly back to the 1638 reformation and, with the same logic of history, back to 1560 and John Knox. The conventicles themselves were, in a real sense, revised versions of the emotional revivals of the 1620s and 1630s, reconstructed to meet the new threats of the 1660s, 1670s, and 1680s.

Although they spread quickly eastward through central Scotland as far as Edinburgh, the conventicles did originate in the west. The very fact of their graduation from houses to barns to fields gave some indication of their enormous size. Descriptions throughout the period listed hundreds or thousands as the expected or witnessed number of hearers. Reporting fast days held at Bothwell, Cambuslang, Mearns, Eastwood, Kilpatrick, and Glasgow—two at each of the last two cities—the Bishop of Glasgow recorded at each fast day an attendance of two or three thousand.[30] Multitudes were reported frequently by both the conventiclers and their opponents. One minister reported that he had baptized twenty or more infants saved by their parents for the offices of the itinerants.[31]

The meetings were large and their functions often sacramental. Baptisms and marriages were performed in the fields despite the prohibitions and legal prescriptions and despite the easy availability of licensed clergy. The conventiclers continued to ordain new, young probationers into the ministry, an action that certainly hinted at a missionary mentality, since Presbyterians, particularly the Scots, did not generally encourage the ordination of ministers-at-large, that is, without a congregational charge.

The Presbyterians also continued the tradition of massive communions, to the dismay of the bishops:

> I am not without apprehension yt if that Conventicle at Austruther be not [retired], seeing that it was so open and resorted to . . . this place of the Country will quickly be pested with such meetings, and people will be debauched from regard to order or church authority, especially seeing at this tyme of the year I am told, yt the ministers whom I have tollerated in this distress notwithstanding of their disobedience to the Law, are preparing to give ther comunions, and thereby proclaim a rendevouz for all the [convenanted] and disorderly people throughout the Country to flock to them as to a banner of separation.[32]

More than even the prayer book sacraments of baptism and communion, these conventicles provided a ritual purification of the community, a congregational (if such can be used when no such legally designated group is meant) conversion. During this period there was far more at stake than merely many sinful individuals who needed to repent their sins so that the community might not defile the Lord's Supper. The communities themselves were, as units, condemned as traitors to the Knoxian tradition. This judgment was not aimed merely at individual congregations or even presbyteries, but at Scotland as a nation. Thus the preachers called for the nation as a whole to repent and return to God.

Within this period the concept of a national conversion and return to Presbyterian tradition clearly paralleled an expected, concrete purification of the country. Certainly the purified community was a symbolic concept. As members of a community recanted their sins and returned to God, their spirits were cleansed and the community purified of the wickedness of its members. During the Restoration period, however, the Scots could actually see impure elements within their nation—the English and their ways, both political and religious. As Archbishop Sharp observed, the conventiclers boasted "nothing to commend themselves to the unstable multitudes who gadde after them, but impudency in railing against the King and publick sentiment, and crying up the Covenant and the good old ranters."[33] Thus the Scottish commitment to national identity and the Scots' intense dislike and distrust of the English were joined to the Presbyterian cause of community conversion. From this union, efforts which might easily be characterized as political took on a deep, religious significance, while the Presbyterian tradition was an acknowledged part of Scotland's independent identity. The political and the religious causes were united, with each side gaining greater strength and legitimacy from this union.

In acknowledgment of this new identification, individual preachers and congregants were expected to recant their sinful behavior of refusing to adhere to and support the Scottish nation and Presbyterianism in the face of persecution. In the earliest stages of these field meetings, the sermon content was thus reported to the Anglican church officials:

> Of late a great many (who for these two years past were silent) have begun to preach and draw after them in some places hundreths, in others thousands, to the open fields; and in their first sermons did acknowledge, and aggravate their sinfull silence, and have promised hereafter to continue in the exercise of their calling even to the hazard of their lives.[34]

They were strengthened in a conviction of their own righteousness and viewed their preachers as prophets calling for the reform of a nation.

> The causes assigned for their fasts are (as we heare) prophanity, breach of covenant, persecution of the people of God. . . . They are of late so numerous and insolent that neither our magistrates nor the few Soldiers which are left here dare meddle with them . . . their greatest encouragement is from the many reiterative assurances with which their evangelists give them that their deliverance and our destruction is neare at hand. . . .[35]

Michael Bruce was undoubtedly one of the best known of the "young men" stirring up trouble through the conventicles in Scotland. Patrick Adair had judged him the "most gifted."[36] Bruce's sermon *The Rattling of the Dry Bones,* preached at Carluke in May 1672, was a masterpiece of rhetoric that demonstrated the full intent and method of these preachers. His choice of scriptural text, Ezekiel's valley of dry bones, suggests the area covered by the sermon. The ten-verse story recounts the tale of the prophet Ezekiel as he was led by the spirit of God to stand before a vale filled with dry bones. God commanded Ezekiel to prophesy to the bones, and when Ezekiel did as commanded, the dry bones came together and became covered with sinew and flesh. As the prophet continued to speak he addressed the four winds, and the breath of God entered the new men so that they "stood upon their feet, an exceeding great army." Bruce, however, did not preach from that concluding text; rather he selected two middle verses:

> So I prophesied as I was commanded: and as I prophesied, there was a noise, and behold a shaking, and the bones came together, bone to his bone. And when I beheld, lo, the sinews and flesh came up upon them, and the skin covered them above: but there was no breath in them.[37]

Aside from the glorious images and auditory constructions available to the preacher using such a text—"bone to his bone," sinews and flesh,

breath, and the shaking—the content of these two verses was particularly
suited to Bruce's purpose. He had chosen a verse pair that began with a
self-justification. He perceived himself as a prophet to the Scottish people,
and therefore his actions were elevated as commanded by God. His selection of the valley of dry bones as the biblical type for the church was
metaphorically and rhetorically sound, for the bones were, when joined
together, the bones of individual persons. The selection of texts from the
middle of the narrative allowed full play for his conviction that the work of
reformation was in progress, but that his hearers would have to undergo
serious changes before they would be prepared to experience the final
ecstasy of conversion, the breath of God's spirit. Through his text, Bruce
was at once able to rail at individual sinners and to point up the dead
condition of the Scottish church.

No published version of a sermon could accurately portray the style of
delivery, and yet the structure of this sermon, the imagistic and rhetorical
devices used, provide some good hints on the actual performance intentions. Beyond his use of crying interjections ("O Carluke!" was called at
regular intervals throughout the text), Bruce had condensed two dual constructions into a single, tripartite device. He employed a dichotomous
structure that took full advantage of the biblical metaphor. He would first
assert a truth or ask a question and then rephrase it within the biblical
terminology. For example: "*O Carluke!* are ye upon the bettering? *O
Carluke!* are ye going bone to bone yet?" As Bruce was following two
themes in the sermon, individual sin and the national or church sin, he
combined the two behind a single, emphatic metaphorical statement.

> What sense have ye of your own sad and dolefull Condition? and of the Sad
> Case and Condition of the Kirk of GOD, and of the sin that hath brought Sad
> Judgments on you both? . . . Is there any *noise* or *shaking* among your Dry
> Bones?[38]

Thus was he able to render the sins of individuals, condemned by all,
equivalent to the current state of the church, and to use the emotional
energy aroused in the cause of each to support the other. Sinners converted
for the good of Scotland, and they fought for the Church of Scotland for the
good of their own souls.

Bruce and his colleagues were not content with their popular follow-
ings in Scotland. These itinerants, especially the very young, went to
Ireland as well to stir the inhabitants against the established Anglican
church. Patrick Adair's *Narrative* recounts the meetings that were held in
solitary places "whither the people in great abundance and with great
alacrity and applause flocked to them." People were contributing to their

support and neglecting their own ministers as "timorous cowards." The itinerants' method of open meetings in public confrontation to the magistrates' authority was thought "great stoutness and gallantry," and "these men were cried up as the only courageous, faithful, and zealous ministers by the common sort of people." The Irish Presbyterian clergy, while convinced of the folly and irresponsibility of these Scottish itinerants, nevertheless could not "disapprove of them lest they lose their people."[39] The enthusiastic response of the Scottish populace was certainly matched in Ulster, but the Irish pastors, hard pressed to protect their ability to serve their people, were less than sympathetic toward the hot-headed, touring preachers who would rile the civil authority and then flee to Scotland whenever times became too difficult.

The major problem was that the Irish Presbyterian ministers, unlike their Scottish colleagues, had successfully carved out quiet pastorates within their province. The deposed Scots had no place to go, no official congregation within which to function; but the Irish Presbyterians did operate within the confines, albeit restrictive, of civil society. Extant records of the presbyteries of Laggan and Antrim, of the congregations of Templepatrick and Burt, tell the story of semisecret, nonpolitical, nondisruptive ministry. These presbyteries, meeting in private houses, continued to examine probationers and candidates for the ministry, to supervise the education of their people, and to oversee pastoral relations—especially the financial issues. Making regular parish visitations throughout this period, the presbyteries would ensure that moral discipline was maintained and the pastor's stipend was paid. Within congregations elders met in session monthly or quarterly, accounting and distributing the collections for the poor, overseeing the minister's house and glebe allocation, and trying to raise the promised stipend. They also held disciplinary hearings, passing judgment and censure upon offenders and absolving those who had done public penance. Most of the crimes brought before them were sexual—fornication or adultery—but the sessions were occasionally needed to arbitrate cases of quarrels, slander, or drunkenness. As long as the congregations and their ministers did not interfere with the magistrates' authority or in any way call attention to themselves as nonconformists, they were permitted to exercise their religion in an uneasy, unofficial toleration. The Irish clergy had decided that their most important duty was to function as pastors, not confront the establishment. They therefore resented the Scottish itinerants who swept through the countryside calling themselves to everyone's attention and leaving the residents to cope with the consequences.

From the restoration of Charles II and the appointment of John

Bramhall as archbishop of Armagh, the Irish Presbyterian clergy quickly learned how to maneuver within a host of severe laws that were only erratically enforced.[40] The most important law for Ireland was the Act of Uniformity, passed in 1662, because it was actually enforced throughout the country. All ministers were required to use the Book of Common Prayer or risk deprivation. Deans, university teachers and students, all pastors, and schoolmasters were to declare openly their conformity to the Anglican church. Any involved in the instruction of youth were required to obtain a license from their bishop, and any clergyman functioning sacramentally was expected to have been ordained episcopally. With these restrictions reinforcing his power, Bramhall deprived sixty-one Presbyterian ministers in Ulster. Nevertheless, any hopes that might have been nurtured regarding the eradication of Presbyterianism in Ireland were soon frustrated by the passive defiance of the ministers. Only eight former Presbyterians conformed to the Church of Ireland. The Church of Ireland also lost the provost, two senior fellows, and five junior fellows of Trinity College. Moreover, six Dublin congregations, two established and four new communities, flocked to the Presbyterian banner.

The Conventicle Act of 1664 and the Five Mile Act of 1665, passed to suppress conventicles and to restrict further the activities of the nonconformists, were also available to the bishops should they want to increase persecution. However, any sharpening of efforts met with serious Irish opposition. The Irish landholders were generally sympathetic to the Presbyterian cause and protected their Presbyterian tenants as far as possible. Only one landlord family, Lord Massereene's, was actually Presbyterian; but most estates were largely populated by Scots-Irish, and the landlords' own interests encouraged them to keep their tenants happy. Indeed, short memories could remember that in 1641 the entire native Irish, Roman Catholic population had risen up in rebellion against the Protestant English, destroying property and lives wherever the rebels fought. The landholders needed the Scots-Irish not merely to farm the land (any tenant could produce income) but to protect their property from insurrection. So the Anglo-Irish elite intervened with the magistrates when necessary, hid the Presbyterian activities from the civil authority, and not only allowed but funded the practice of Presbyterianism on their estates. In some instances the landlords provided house and glebe for a minister; in others they actually built a church edifice for the Presbyterian tenants.

The year 1672 brought additional relief and pressure upon Irish Presbyterians. The Test Act was passed, requiring anyone holding office under the British government to take an Oath of Supremacy and Allegiance, not only to the king, but to the Church of England. Also at this time, Charles was made sensible of the debt he actually owed the Irish Presbyterians who had supported his cause during the Commonwealth period. "Unhappy Presbyterians suffering for their loyalty to the King and now suffering under him."[41] The king made the Presbyterian ministers a royal grant—a regium donum—of four hundred pounds annually (later increased to six hundred) to be divided among the clergy. Meant to offset the inadequate compensation suffered by pastors due to the poverty and tithing demanded of their people, the regium donum enabled ministers to survive economically while working full-time as pastors. The king's recognition and financial support only exacerbated the anger and determination of the official Church of Ireland, and the bishops resolved to employ the tools of persecution already at their disposal.

There began a rash of writs and summonses for illegal marriages, nonattendance at services, and frequenting of conventicles. In 1676 a correspondent from the Presbytery of Route told his colleagues at Laggan that at least eight score men from various congregations had been summoned to courts of excommunication, some forced to pay large fines and some threatened with prison. Again in 1678 the same trouble was reported in Dublin, Lifford, and Longfield. Some clergymen were fined for lack of a written certificate of episcopal ordination; presbyteries were searched out and their members harassed. The Antrim and Down meetinghouses were closed down by force in 1684, and the public services of Presbyterians were forbidden.[42]

In addition to keeping their work as quiet and hidden as possible, the Presbyterian clergy had developed several strategies for coping with the persecution. After the Laggan Presbytery had ordained someone to the ministry, the candidate was immediately sent to Scotland for an undetermined period of time, usually a few months, before he would be expected to return and take up his charge in Ulster. This gave the appearance of ordained clergy coming from Scotland, and protected the presbyteries from the charge of illegally ordaining ministers of the gospel. Congregations developed networks with other meetings to guide invited Scottish itinerants through the province so that their presence would go more or less unnoticed by the government. They also provided refuges and hiding places for each other as the bishops in one area clamped down while others allowed congregations to meet mostly unnoticed.

Because this organization was necessarily unofficial and sporadic, the clergy proved unable to thoroughly control Presbyterian conduct. The presence of Michael Bruce and his colleagues stirred up the laity, undermined Irish clerical authority, and created problems with the civil authority; but the presbyteries had no means by which to control the progress or activities of these preachers. The congregation of Sligo petitioned the Laggan Meeting in 1672 that William Paterson, who was preaching and baptizing among them, be ordained and continued there. The presbytery responded by warning Sligo of the "sinfullness and danger of William Patersons practices of usurping the ministeriall calling without the tryall of his gifts and Ordination and that therefore they ought to avoid him as ane intruder."[43] Their reputation for fractiousness and the disruptive influence of the Scots were so well known as to be discussed among the king's advisers. In one letter to the Viceroy Ormonde, the Lord Chancellor Boyle reported:

> [that] they in the North are a numerous rabble is very well known to your Grace, and that they will never want [for] discontented factionists to inflame them upon all overtures of trouble cannot be much doubted while they are so near in neighborhood of Scotland—and that the most violent, most discontented of that kingdom have the freedom of coming over hither when they please. . . .[44]

When the bishop of Killyleagh and Achoury complained to the Earl of Essex, it was not about the Irish church members but the itinerants that he wrote, petitioning for

> the few Protestants of these parts that they may be saved from Scotch presbyters who ramble up & down to debauch the people in their religion and loyalty. . . . Two of these Genevan calves . . . held forth within two miles of me to the great peril of the apron strings, which were much endangered by the deep sighs of the waistcoateers. . . .[45]

Bound on one side by the strictures of civil government and pressured on the other by the antics of the fearless youths able to flee the consequences of their deeds, the ministers worked to protect the safety and maintain the loyalty of their charges.

Throughout this period the ministers continued to exercise their sacramental functions and, in part, maintain the attention and support of the people. Communions were celebrated on a surprisingly large scale, considering the observation under which the congregations were kept. When asked by Robert Campbell whether he might "freely & safely" set about giving communion in Raigh "in as publick a manner as before," the Laggan Meeting recommended that he go ahead "as he had done before," but that he not wait until after the harvest, since that would give time for news to spread. In Ramelton, a full three-day communion service was scheduled in August, with four preachers listed to speak during those services.[46]

The sessions records of the congregation of Burt from 1678 to 1682 note an annual celebration of the Lord's Supper. These celebrations followed the standard timetable commonly observed since 1630 of a fast day preceding the event, preaching all day Saturday, distribution of the sacrament all day Sunday, and thanksgiving on Monday. Two or three ministers, plus the local pastor, William Hempton, were always invited to help, and the preaching was divided equally among all clerical participants. In 1680 the congregation did not sponsor a communion service, but instead was encouraged to go to Raigh, where its pastor was assisting. In 1681, on the other hand, they sponsored an extraordinarily large service, for Ireland, with a day of humiliation led by two visiting ministers, a three-day service that included six clerics, and an attendance so large that two preachers spoke to the crowds outside the church while individuals inside the building received the sacrament. During one season, as if the communion were not enough, baptisms were held the Monday after.[47] One Church of Ireland member wrote scornfully that ministers and elders

> have an absolute government over their congregations; and at their Communions they often meet from several districts to the number of four or five thousand. Then they preach in the fields and continue a great part of the day together and think themselves so formidable that no government can touch them.[48]

Clearly, then, Irish Presbyterians during the Restoration period participated in two distinct variants of ritual congregational behavior. The Irish ministers encouraged a continuation of the services or practices begun in the late 1620s, when Edward Bryce, Robert Blair, and others had functioned within, although not actually as a part of, the prelatic Church of Ireland. Worship was organized around openly sacramental occasions, usually communions, but occasionally, in this later period, baptisms. These services were still multiday affairs that involved hundreds, if not thousands, of people. How raucous these events were is not known, although it is unlikely that ministers who were maintaining a delicately balanced toleration would have risked that tolerance by encouraging loud, boisterous, emotional expressions of faith and religiosity. The size and length of the events, if not the actual activity, were enough to bring some bishops to prosecute Presbyterians, both ministers and lay persons, who were forced frequently to seek the protection of their English landlords.

Yet despite this testimonial, many among the laity were dissatisfied with these services as ineffectual and weak. The people either could not comprehend or would not sympathize with their presbyteries' efforts to remain and work quietly in Ireland. The laity saw only the ministers' unwillingness to challenge English authority and explained it as fear for their lives and livelihoods.

This perception of ministerial silence as cowardice was encouraged, if not ignited, by the activities of the young, Scottish itinerants. The traveling conventicles of these itinerants provided the other type of ritual available to the Presbyterian laity, and, much to the chagrin of their pastors, vast numbers of people flocked to hear the traveling preachers. These meetings are known to have been highly emotional, with the preachers deliberately provoking zealous response and active participation of the laity. Moreover, the content of the sermons was quite similar to that of the sermons delivered in Scotland, that is, the unjust dominion of the English and the Anglican establishment over the Scots and Scottish institutions, and the spiritual duty of all Scots to resist English authority. In this framework the Irish Presbyterian ministers were condemned for not arousing the Anglican ire and for seeking or enjoying the protection of a few powerful Englishmen. The Irish needed not only to purify their neighborhoods from English influence but also to purify themselves from the heinous sin of collaboration with the English. The mild, symbolic purification that the community experienced during sacramental occasions paled, for many, in the face of an overriding political/religious mandate. Thus the years of the reign of Charles II represented the first period during which the Presbyterian leadership was at odds with its community and the ministers' carefully considered actions were challenged by the laity.

Still, the community survived even through the brief, intense persecution of the early 1680s. James II ascended the throne in 1685 and in 1687 passed his Declaration of Indulgence for dissenters and papists. Although these acts provided relief for nonconformist Protestants, James' real goal of abetting the Catholics was soon apparent. This realization brought with it the Glorious, "bloodless," Revolution in England, which in turn brought a small war in Ireland. During the 1680s the Stuart king had been gradually augmenting his forces by occupying and exploiting the resources of Irish towns and villages, but the actual military campaign was brief and decisive. In mid-June 1690 William III landed at Carrickfergus with a small army of about thirty thousand and began a march to Dublin. On 1 July he defeated James at the Battle of the Boyne, with James seeking refuge in France and William returning to England. The native Irish were clearly behind the Stuart standard, while the Presbyterians and, after careful thought, the Anglicans supported William. In response to the Protestant support the monarch could not justly disestablish the Church of Ireland, nor could he convince the Irish parliament to ratify the provisions for the toleration of the Presbyterians in the Treaty of Limerick. Nonetheless, he did continue the regium donum, doubling the gift to twelve hundred pounds per year, and he successfully moved Parliament to repeal the Oath of Supremacy in Ireland. Now the Irish Presbyterian church was in the peculiar position of being concretely supported by the king, with its members permitted to hold local civil offices, while its ministers and congregants could still suffer occasional persecution for their participation in Presbyterian religious services.

Although the fate of the Irish Presbyterians would be subject to the whims of the British monarchy for another thirty years, their troubles were essentially over. Unable to be the established church, they were content with a tolerated, dissenter status, partially supported by the government, but still dependent upon a poor, economically overburdened population. The ministers regrouped their network, divided anew presbyterial and synodical assignments, and met for the first time in thirty years in general synod. The General Synod of Ulster was from this point a national organization, separate from the Scottish church. While acknowledging the Church of Scotland as its forebear and frequently seeking Scottish education and advice, the Irish synod refused to be bound by any Scottish decisions.

For thirty years the two national communities had experienced different sorts of hostile, civil pressure. The Scots had been forcibly denied their Presbyterianism after having had some enjoyment of it for twenty years. Defiance and violent rebellion were their answer to the prelatic establishment when no loopholes were provided for the extremists. The moderates, themselves willing to conform, seemed unwilling to harm their less flexible countrymen. The Scots-Irish had been allowed some flexibility through most of that period, the same kind of toleration balance that had characterized the early decades of the century. They responded by taking care to keep opportunities available so that they could minister. Trying desperately to enable the continued operation of the church, the clergy righteously resented the activities of the Scottish itinerants, not only because the itinerants brought down the wrath of the civil authority and made their jobs difficult, but also because the itinerants alienated congregations from their pastors.

These initial years of stability demonstrated that while two distinct tendencies existed in these communities, one toward the grand, emotional, spiritual experience and the other toward quiet study and performing one's moral obligations, the line was not drawn geographically but vocationally. During those years of tribulation, the Irish laity supported the extroverted activities of unattached missionaries who were destined to inflame a tense situation. The people seemed happiest not when a minister visited and privately catechized their children, but when a preacher held a large, liturgical service. Moreover, in Scotland, after the Glorious Revolution, the clergy strove to quell the excitement, in effect to pick up where they had left off in 1590 before the Stuarts started the religious disputes. Indeed, when reconstituting the church many of the 1590 codes were revived and inserted while the 1638 reformation and the activities of the Remonstrators and the zealots of the previous three decades were hardly mentioned.

Although the period of political and religious protest had passed, the clerical-lay stress was still apparent in the continued discussion of communion services. The twenty years following the establishment of the Ulster Synod saw an increase in the number of celebrations as well as the regular scheduling of annual events. In Burt a communion service was held in 1694 that required five ministers and thirty-six bottles of wine. The description of a communion at Burt two years later suggested that the congregation had not lost sight of the conversion that the community was expected to experience. Elders were still required to visit every communicant beforehand to ensure fitness and to resolve any scandals before that Saturday.[49] The Laggan and Route meetings still needed to assign two or three ministers to assist at services far away from the generality of congregations. At Templepatrick the communion service required the purchase of ten gallons of wine and four dozen loaves, the use of three flagons, two bowls, and four cups, and the services of sixteen elders to aid at least three ministers in distributing the elements and collecting the money. One congregation in the Presbytery of Monaghan had so many communicants that the church had to use unordained laymen because they lacked sufficient elders, much to the horror of the surrounding congregations.[50] And people clearly felt called to participate frequently in these services. As the Session of Larne recorded:

> This day it was intimated that we will not have Sermon next Lord's Day because Mr. Ogilvie will be at Carncastle Communion and people exhorted to repair thither for hearing of the Word and communicate (as many) as may be admitted and are willing.[51]

The awkward point was that the official church had begun to discourage these large events. In an effort to restrict size, the Ulster synod decided that

> at the Celebration of the Lord's Supper there be only 3 Ministers employ'd with the Minister of the place, & for avoiding great Multitudes that resort to Communions, that the neighboring Ministers be not call'd at such Times unless it be on Saturday or Munday.[52]

Yet it seemed that the assembly objected to more than size; at the same session the synod passed a resolution "that Ministers & Preachers use a sound Form of Words in Preaching, abstaining from all romantick Expressions & hard Words, which the Vulgar do not understand, as also from all Sordid Words and Phrases."[53]

As might have been expected, considering the religious situation in Scotland before the Glorious Revolution, the Scottish church had greater difficulty controlling its people. The Scottish congregations had, in essence, experienced total overhauls in 1687, with the majority of communities acquiring new ministers, and elders attending newly organized presbyteries. The ousting of prelacy notwithstanding, many of those Anglican Scots were permitted to conform to Presbyterianism, and they were granted new congregations upon condition of free demission of their previous charges, subjection to the Presbyterian government, and allegiance to William and Mary. While the extremists condemned (and for a century thereafter continued to condemn) this leniency, the new clerical majority was devoted to peace. The people did not seem to object to peace; what they disliked was quiet.

In 1692 the Synod of Glasgow and Ayr began a doomed effort to restrict the communions:

> [A]ll the brethren within the bounds of this Synod [may] have as frequent communions as may be, and . . . invite as few brethren to help them thereat as they can, that so all confusions and disorder may be prevented.[54]

Apparently forgetting that people anxiously attended such events eight or ten times annually, this synod again sought in 1694 to curtail communions by requiring every parish to hold at least one annually. "For preventing prophanation of the Lord's day at Communions by great confluences of people," in 1697 the synod restated the once-a-year requirement and recommended that only three or four ministers be invited to assist and, further, that no minister come unless invited. Ten years later the problem still haunted the area, bringing the additional demands that communion be distributed twice annually in towns, that ministers from the surrounding neighborhood not be invited, that all pastors present certificates testifying to one or two services a year, and that presbyteries supervise pastors and set up schedules.[55] The synod probably would have done well to accept the inevitable as did the Presbytery of Paisley:

> Anent the transmitted overture about the Regulating Confluences of people at Communions judges it impracticable as it stands but thinks it should be Recommended to all Ministers to take prudent and Convenient measures for

the Preventing the Prophanation of the Lords day by too great Confluences.[56]

In some ways the Restoration period was an interruption in the development of the Presbyterian church. Certainly there could be no institutional
development as such, because for almost thirty years the Presbyterian
organization had gone underground while an episcopal structure was
imposed upon the nation. The conflicts of this period had much more to do
with politics and national identity than with religion and church polity. Yet
the conventiclers invoked God, not Scotland; their avowed goal was spiritual integrity, not the reaffirmation of Scottish nationalism; and they discussed the need to purify the community, not oust the English. Even when
they openly described the English presence as an evil and encouraged the
people to resist, the preachers focused upon the Anglican church and the
pollution of the Knoxian tradition. On the other hand, Presbyterianism had
almost become another facet of the Scottish political identity. Political
circumstances so influenced religion and church structures so informed
political perceptions that religion and politics became convoluted in a
single protest process. Religious conventicles were havens of political
unrest, and political dissidents were brought to attend religious meetings as
a primary outlet and organizing force for their anger.

In Scotland, the nonconforming ministers, whether consciously or
not, used the political threat to enhance their control, power, and influence
over the people. With the civil establishment in Scotland aligned with the
Stuart interest, the dissenting clergy became national Scottish heroes, the
only leaders protesting the English presence. They used religious rituals
rooted in the revivals of the previous decades and manipulated their structure to better fit the contemporary exigencies. Through their effort revivalist meetings, which had before 1650 been focused in the western counties,
spread throughout the Lowlands, even to the city of Edinburgh itself.
Unable to exist at all without popular support, the conventiclers organized
their lay followers in support and fomented staunch resistance—in some
cases armed resistance—to their opponents.

In Ireland the religious situation was more complex due to the ambig-
uous status of the Presbyterian church there. The ministers retained an
authority from the uninterrupted progress, since 1642, of the dissenting
Protestant organization. They continued the popular rituals of the earlier
period, though with less flamboyance, and experienced some pastoral
success until challenged by the politically extremist itinerants from Scotland. The irrelevance of the Scottish position to the Irish political situation
did not stop the Irish laity from identifying with Scotland and flocking to

the more radical standard. Lay sympathy went with the Scottish itinerants, with the goal of a pure, converted community far outdistancing, in their minds, the less romantic sins of continued sufferance and unofficial toleration. Additionally, with this ideological priority went a decided preference for community revivalist forms and rituals as opposed to quieter, sometimes private, nondisruptive exercises encouraged by the beleaguered Irish clergy.

When in 1690 the persecution was essentially ended, the people and their pastors were destined to drift further apart. Free from civil interference, the ministers would be able to pursue their theological studies and church expansion in peace. Now guaranteed of financial support, fully in Scotland and partially in Ulster, now responsible for congregations in an increasingly stable society, and now guaranteed pastoral control over their congregations, the clergy felt able to devote themselves to theology and institutional integrity. The conventicles had served their purpose, and the time had come for the clergy to leave behind those deviant practices, necessitated by political circumstances, and turn to more serious, intellectual and devotional activities. The people, nurtured in the unstable climate of the seventeenth century, sought continued emphasis on the spiritual journey as a community event calling forth great emotional involvement and commitment on the part of the individual. The progress of the Enlightenment was giving educated ministers new hope in the intellect; the traditions of the conventicles firmly grounded the people in zeal. In Lanark in 1702 the clergy complained that

> the ministers of the gospell particularly in those bounds have been exposed to the mistakes of many weak though otherways well-meaning people, as if they did not maintain a just zeal for true presbyterian principles.

As the eighteenth century continued, the enlightened ministers would find it increasingly difficult to convince "that misled people and the world" of their "constancie to the true principles of the covenanted work of reformation in this land."[57]

3

Subscription and the Power of Orthodoxy

For the Presbyterians of Ireland and Scotland, the reign of William and Mary was a joyous respite from the years of political pressure that they had endured during the final years of the Stuarts. Perceiving himself as indebted to his Presbyterian subjects in those two countries, William rewarded the Presbyterian Church of Scotland with official recognition as the state church and the Irish Presbyterians with unofficial toleration. Under the patronage of a sympathetic Calvinist monarch, the Presbyterian clergy of both Ireland and Scotland were able to increase the number of congregations, expand the church's moral influence, and stabilize the internal administration. Once again congregations exercised discipline in an orderly fashion, presbyteries addressed themselves to the problems of educating their ministry, and the Ulster Synod and General Assembly of Scotland became burdened with keeping peace among incorrigible factions within the churches.

Unfortunately, the peace with the English government was short-lived. Queen Anne Stuart succeeded William as monarch, and Anne, high Tory like her father James, was a sympathetic listener to the advice of her conservative Anglican advisors. Every year the Irish clergy had to battle anew for the payment of the regium donum; in 1714 the Tory party finally convinced Anne to cease payment altogether. When George I succeeded to the throne, he immediately reinstated the clerical grant, being himself a Calvinist; but he was unable to enact official toleration for the Irish Presbyterians. Yet, despite the passage of the Test Act, which had effectively disfranchised the Presbyterians in Ireland, and despite sporadic persecution by the Anglican establishment, the Irish Presbyterians found themselves free to argue among themselves. In fact, the initial efforts to seek inclusion in an official Act of Toleration sparked a major battle in the Irish church: the subscription controversy.

The actual disagreement concerned the desirability of subscription to an established creed as a reasonable guarantor of consensus and unity. The creed in question was the Westminster Confession, with its accompanying catechisms and books of discipline. No one challenged the validity of the confession itself; the Westminster Confession was perceived by all participants as the most perfect of creeds. The problem, as discussed publicly, focused upon the act of subscribing any creed as a test of orthodoxy. However, the avidity with which both sides pursued their own positions, to the point of schism, appeared inconsistent with this focus. Why was it not enough that all agreed upon the Westminster Confession as the best expression of the Christian faith?

This controversy was quite complicated, for it involved the response of the Ulster community to three separate Presbyterian networks: the organizations of Scotland, England, and southern Ireland. Thus it is necessary to trace events in all four communities in order to understand how the General Synod of Ulster, united and peaceful in 1710, could be led to schism over this issue in less than twenty years. Moreover, it should be noted that subscription was not a simple point of theological debate argued by the religious elite. The laity in all four regions, but especially in Ireland, actively participated in the various conflicts, and this relatively minor issue of polity was built up into a question of theological orthodoxy and spiritual integrity. Therefore, not only must the intellectual debate be considered, but also the problems of church politics and clerical-lay relations. When all these factors have been assembled, it will be clear why, for a brief moment, the most important issues facing the Ulster Presbyterians were reduced to the question of subscription to the Westminster Confession.

The Westminster Confession itself had become part of the Knoxian tradition. Linked as they were to the 1638 reformation and the Solemn League and Covenant, the Westminster documents had superseded Knox's writings as the primary statements of beliefs and practices. All through the latter half of the seventeenth century, the Presbyterian ministers in northern Ireland, doing their best to educate and care for their people privately, had used the Shorter and Longer Catechisms as their major catechizing tools. Some ministers required parents' assent to the Westminster Confession before baptizing their children; most presbyteries required a candidate's assent before approving his ordination. The Irish laity therefore grew familiar with the confession, not so much in understanding and evaluating specific tenets as in recognizing the language and rhetoric and perceiving it as the true Presbyterian creed. In this sense the Westminster Confession was the definitive summary of the Knoxian tradition, the primary theological formulation. Any challenge to the confession represented a threat to the entire system of belief, polity, and ritual practices. The Westminster Confession was the credal guardian of true religion, an easily identified symbol of Scottish Presbyterian orthodoxy.

At the beginning of the eighteenth century, a few theologians had embarked upon paths of intellectual exploration that could only be evaluated as heretical by most contemporary Christians. Influenced by the Enlightenment's emphasis upon in-depth intellectual investigation and the new impulse to examine heretofore unquestioned assumptions, some theologians were challenging doctrines that had never before been disputed. The most threatening challenge involved efforts to redefine the nature of the Trinity, a doctrine that had always been accepted without doubt by all Christians. Early in the eighteenth century, however, one man who had openly denied Trinitarian dogma was tried by church tribunals, bringing the insidious threat of this new heresy to the attention of the entire British Christian community.

The trial involved the Southern Association, a thriving Presbyterian network in the south of Ireland, in communion with, though distinct from, the Ulster Synod, rooted not in Scottish Presbyterianism but in English nonconformity. The case concerned Thomas Emlyn, graduate of Emmanuel College at Cambridge, assistant to Joseph Boyse of the Wood Street Church, the largest Presbyterian congregation in Dublin. Having read William Sherlock's latest treatise on the Trinity, Emlyn developed some peculiar notions. He came to affirm the subordinate theory of God that became identified with the Arians, namely, that only God the Father was supreme, Jesus Christ the Son was divine only as divinity was communicable from the Father, and the Holy Spirit was inferior to both the Father and Son in order, dominion, and authority. Though Emlyn affirmed the divinity and the preexistence of the Son (in opposition to the more secular Socinians who accepted the virgin birth and the resurrection but rejected the essential divinity of Christ), he was far from agreeable to orthodox Protestants who found in the doctrine of the Trinity no room for skepticism. In response to his preaching his new beliefs, Emlyn in 1702 was presbyterially tried and excluded from the communion at Dublin. He removed to England where he expected to find fellow believers within the Anglican church.[1]

The Trinity had always been a safe, undisputed doctrine, one that Roman Catholics and Protestants agreed upon. Questioning the divinity of Christ undermined the very essence of Christianity. But during the last quarter of the seventeenth century, a few English theologians had begun to open up this debate. Pamphlets had been hurled back and forth as the old heresies of Arius and Socinus were revived and explored. Yet before 1700 the debate had remained intellectual, generally affecting only the polemicists. As the century turned, the issues filtered down to popular levels, and individual clerics like Thomas Emlyn were cast out of their pulpits. Circumstances were not helped by the publication, in 1712, of Samuel Clarke's *Scripture Doctrine of the Trinity*. Here was no eccentric intellectual, but a serious thinker challenging the established doctrine of the Trinity. While Presbyterians in Scotland and Ulster could easily blame the flagrant Church of England for her own problems, Protestants in England, nonconformist as well as Anglican, were seriously disturbed.

Fears of heterodoxy now began to haunt congregations as well as certain zealous ministers. The dissenting congregation at Stoke-Newington, turned upon their pastor, Martin Tomkins, demanding that he prove himself orthodox on the Trinity. Tomkins refused to acquiesce to the congregational demand. He claimed that only Christ could establish terms of communion, and that no one, no matter what error was suspected, could be judged without reference to scriptural injunction. In his defending pamphlet, Tomkins loudly proclaimed certain principles of scriptural freedom; but he also provided a lengthy discussion of Arian doctrine, along with scriptural justifications for it. While he repeatedly asserted that he had never espoused Arianism, he did admit that he had tried to demonstrate that Arianism was not unbiblical. The congregation refused to accept him without an open declaration, Tomkins refused on principle to give them such satisfaction, and the people removed him from his pulpit.[2]

Of greater import was the incident in Exeter that began with the questioning of students at the nearby Calvinist academy and ended with the congregation's ejecting James Pierce and his colleagues from the pastorate. Joseph Hallett, a young student, had begun to discuss openly and eagerly the controversial works assigned to the students, including Clarke's treatise on the Trinity. Accusations against the instructors clearly involved no breach of morality or any other point of dogma. Pierce was charged with "using great Violence and Artifice in carrying on his Attempts against the ever blessed *Trinity* and making a very strange Progress in perverting the *Youth* of that City."[3] The congregation called upon seven neighboring ministers for advice, and the seven agreed to interfere. The conference eventually demanded that Pierce and his colleagues prove their orthodoxy by subscribing the first of the Thirty-Nine Articles of the Church of England—not an overwhelming demand, since all Protestant ministers in England had to subscribe this article in order to ensure their own religious liberty.[4] Pierce and company refused to subscribe, and the proprietor of the church barred the door. The congregation sided with the proprietor; they attended debates, reaffirmed the decision, declared their satisfaction, and brought upon themselves Pierce's public contempt of lay involvement in clergy matters.[5]

From this flurry of activity, it was only a brief step to a general convocation. In March 1719, at Salters' Hall in London, the English nonconformist ministers were invited to participate in a discussion of the rights of congregations, ministers, and clerical courts. The sessions were attended by dissenting clergy who identified themselves as Presbyterians, Independents, or Baptists. No laity attended; that is, no lay members had floor or voting privileges, though some did attend and observe. It is difficult to know how much of the distrust and disunity at this convocation resulted from immediate anxiety concerning the defections within the Church of England and how much was rooted in the old competition between the Presbyterians and the Independents. The alignment of ministers on this issue to some extent followed the old lines. For example, Drysdale, in his *History of the Presbyterians in England,* notes that two thirds of the Presbyterians were nonsubscribers, while two thirds of the Independents favored subscription.[6] Still the changes of the decade must have sparked this particular debate. This battle was quick and sharp, with both sides shouting at each other on the floor as well as self-righteously exercising themselves through the printing industry of London.

On one side were those who feared Arianism, Socinianism, and any other error contradicting Trinitarian dogma. They believed that a "Subscription of such *Humane Words*" was necessary to ensure the purity of the church and to help convict those in error. To demonstrate their commitment, they subscribed publicly the first of the Thirty-Nine Articles and the fifth and sixth questions of the Westminster Catechism. Their opponents dreaded lest "any Breach should be made on Justice, Charity, or Christian Liberty, in opposing *Arian,* or other Errors, which usually gives Advantage to the Erroneous, and is a Scandal and Damage to Christianity." While accepting the desirability of public declaration, the proponents of liberty felt that a statement in "Words of Scripture" would be safer and more effective; they declared themselves against Arianism and in favor of orthodox doctrine on the Trinity because such was grounded in the Bible. Finally, they refused to oppose any individual who declared his beliefs in scripture words and refused to subscribe any human construction.[7]

Those opposed to subscription were fighting not only for Christian liberty, as they called it, but also for clerical power. In one anonymous, antisubscription pamphlet, the congregation at Stoke-Newington was accused of fomenting the distress that resulted in Tomkins's removal. The author also described a congregation that had excluded its ministers for refusing to subscribe a set of articles that the congregation itself had drawn up. The entire problem was summarized in the question attributed to a clerical delegate at the convocation: "Whether that Assembly was to be directed by the laity?"[8] Another anonymous antisubscriber was undoubtedly sincere when he said,

> I love Liberty and the Scriptures; and hate a Prison for my Body or Creeds for my Mind; and cannot but acknowledge I dislike to have anything impos'd upon me from an Assembly of Divines (so called) as much as *Luke Melbourne's* Antagonist professes to disapprove any thing offered by the Laity.[9]

The preoccupation of these pamphlets with lay interference indicates that the clergy was experiencing a real threat to its authority.

The nonsubscribers had seen the urgent congregational demands for reassurance upon their orthodoxy as an inquisitorial effort to constrain their religious freedom. Yet they were not the only ones to see dragons. Thomas Bradbury, a leader of the prosubscription party, accused his opponents of poisoning the purity of the church, not merely by making it possible for heretics to slip in, but by actually trying to bring in the errors themselves. If they were willing to state their beliefs in their own words, he asked, why were they not willing to subscribe a formula? Surely they did not believe their own phrases to be closer to the revealed word of God, for their sentences did not always come from the Bible. Furthermore, Bradbury believed that he or any congregation had the right to understand a pastor's declaration of faith. If a minister accepted Christianity as expressed by the Westminster Assembly, why was he unwilling to subscribe it? For if a man did not accept the confession, he should not be a minister within the Christian church.[10] So continued the arguments, with neither side denying the validity of what its opponents said, but neither side believing that what was said was what was meant.

Despite the vituperative spitting on both sides, the Salters' Hall Convocation did reach several conclusions; unable to expect future unity among the dissenting groups, the majority developed a platform for peaceful coexistence. The final vote was opposed to the requirement of pro forma subscription. However, the assembly did recognize that certain errors might warrant, even oblige, a congregation to withdraw from its pastor; moreover, a congregation had the right to judge its minister. The convocation then proposed advices to be followed by all communicants in order to ensure the continuity of the Protestant church. A congregation with grievances against its pastor should have proof before seeking assistance from other ministers. Only with the failure of private admonition should a public hearing be held; and at a public hearing a man was responsible only to scriptural teachings. Although catechisms and creeds were allowed as great aids by "wise and learned" men, they were also created by "fallible" men. While some acknowledgment of the existence of truth and error was made, and some recognition of congregational rights was granted, the convention's concluding "Advices for Peace" were a firm victory for the nonsubscribers. The convocation went further and officially censured the subscribing ministers when they, on their own initiative, subscribed a declaration of orthodoxy.

> We saw no Reason to think that a *Declaration* in *other Words* than those of *Scripture* would serve the Cause of *Peace* and *Truth;* but rather be the Occasion of *greater Confusions* and *Disorders; . . .* . May not the Method they have taken, tempt men to question, Whether the Scriptures be *so perfect a Rule of Faith,* as Protestants have all along represented them to be?. . . when we betake Ourselves to *Humane Declarations,* and dare not depend upon those which are Unquestionably *Divine?*[11]

The validity of subscription to an established creed was no problem for the Church of Scotland. Having always perceived itself as the national church, even when not sanctioned by the civil authority, the Presbyterian Church of Scotland had generally abided by the kinds of rules that structured a national church, including formal statements on theology, church polity, and moral discipline. From 1647 the Scottish Presbyterians had seen the Westminster Assembly documents as encompassing the best, broadest formulation. Most ministers had subscribed the Westminster Confession or, if they had not, were understood to be in agreement with it.

The problem in Scotland came to be: what was meant by subscription? Did a subscriber accept with the same degree of commitment every tenet of the confession, despite the relative importance of each in the realm of theology? Was it possible that some sections were not necessary to salvation, and, if so, how was subscription with certain exceptions to be understood? The crisis was initiated by the Presbytery of Auchterarder during the ordination trials of probationer William Craig. In 1716, upon receipt of unsatisfactory reports, the presbytery began to question Craig upon specific points of doctrine. Drawing up his answers into a declaration of his comprehension of the Westminster Confession, the presbytery granted Craig a license to preach provided that he subscribe the declaration, which he did. Within a month Craig challenged the paper, demanding scriptural references for certain questionable articles. The presbytery answered that such could be found but they had not the time and were not bound to find them. Craig then rejected the paper, and his license was voided. Craig appealed to the synod, which referred the matter to the General Assembly, which ordered the presbytery to return his license.

The actual problem centered upon the presbytery's assertion that "it is not sound or orthodox to teach, that we must forsake sin in order to our coming to Christ and initiating us in Covenant with God."[12] The presbytery maintained that such a belief compromised the doctrines of free grace and divine omnipotence. In their defense, the Presbytery of Auchterarder explained that they did not mean that it was not one's duty to forsake sin, that any could come to Christ without being sensible and sick of sin, or that awakened sinners did not have convictions to forsake sin. What the presbytery did mean was that the forsaking of sin was not antecedently necessary to the sinner's immediate duty of coming to and believing in Christ. Any other method of instruction would lead sinners to despair and tend toward legalism. Eventually the General Assembly accepted the explanation of the Presbytery of Auchterarder. However, the assembly did warn against the statement in question, lest it lead the less educated lay person into antinomian error. Furthermore, William Craig was licensed upon his own declaration, and Auchterarder was instructed to restrict its questions during ministerial trials to the actual forms of the Bible or the Westminster Confession.[13] After the conclusion of this episode in 1718, no presbytery tried to force exceptional requirements upon its candidates.

Although the crisis in Auchterarder eventually subsided, upheaval in the national church had barely begun. In 1720 was published *A Letter concerning the Defections, Sins, and Backslidings of the Church of Scotland,* an exceedingly brief diatribe (considering the scope) against all the sins plaguing church and state since 1690. In grand Calvinist style the anonymous author provided lists upon lists of evils. On the negative side, the church had neglected to assert boldly the headship of Christ, to declare the intrinsic power of the church to call assemblies and appoint fast days, to return to the glorious days of 1638–49, to be faithful to its own interest and the memory of William III, to declare Presbyterianism of divine institution, to condemn prelacy, and to demand adequate repentance from those ministers who had complied with the toleration during the reign of Charles II. The church had buried "a Covenanted Work of Reformation at the unhappy Union" with England in 1707. The nine accusations that followed this attack, such as "The giving of our Civil Rights and Privileges to a Nation that never kept Faith to us, are too strong for us; and so turning our Nation and State into a Nonentity," would have been more properly aimed at the civil government. Nevertheless, as the Presbyterian church was, in fact, the national establishment, the author held the ecclesiastic authority partly responsible for the incidental consequences of the union, including the reintroduction of plays and balls, boundless toleration, and the multiplication of unnecessary oaths.

Following an identification of the progress of the church with that of the nation, the author was then free to detail the intrachurch sins. Complaints ranged from the very abstract and undemonstrable, such as the pervasiveness of a "legal" spirit and want of zeal, through unorthodox practices, such as the growing prevalence of private baptisms and irregular marriages, to very precise indictments of the conduct of the church's judicatures. The author maligned the assembly's treatment of the Presbytery of Auchterarder, and he disputed the existence and actions of the newly appointed Commission on Doctrine, charging the court with unfairness in the manner in which moderators and commissioners were chosen. Finally, he decried the overpowering influence exercised by the ministers in Edinburgh in opposition to those from central and western Scotland.[14]

In essence, this pamphlet was a complete catalogue of the dissatisfaction that a significant minority of clergymen felt with the growing ascendance of the moderate party over the evangelicals within the Church of Scotland. Struggling for dominance of the Presbyterian church, the two parties were almost personifications of the two competing forces within Calvinism: piety and reason. The evangelical party strove to retain that awesome sense of the infinite power, wisdom, and love of God. They preached that a person must first be open to Christ and receive the grace of the Holy Spirit in order to be saved. Consequently, any who encouraged people to uphold high moral standards as indicators of God's grace were labeled legal preachers, guilty of teaching that salvation could be attained through meritorious efforts and deeds. The moderates emphasized rational theology, preaching that one must by reason learn to control the sinful passions. Although they agreed with the evangelicals that God's saving grace was absolutely freely given, the moderates believed that individuals were obliged to make their conduct pleasing to God. The moderates said that the evangelicals were antinomians; the evangelicals said that the moderates were Arminians.

In an effort to better instruct his congregation, James Hogg in 1718 reissued Edward Fisher's *The Marrow of Modern Divinity*.[15] A treatise on free grace, opposed to any attempt to recognize a role for humanity in the process of salvation, the *Marrow* did contain some problematic expressions. Fisher maintained that all people joined to Christ by the covenant of grace were no longer subject to the covenant of works, that the saved were in no danger from the penalties of the law, and that the actions of the saved had no relevance to the attainment of eternal life since all righteousness was imputed to them through Christ. Anxious over the decline of their congregations into immorality and anarchy, the moderate majority of the General Assembly appointed a Committee on the Purity of Doctrine to study the *Marrow*. Predictably, the committee's report concluded that the *Marrow* was heretical, specifically antinomian, and advised that pastors instruct their congregants to avoid such erroneous works. This report was quickly approved by the full assembly.

The outpouring of pamphlets that followed this decision involved two different, but connected, issues. The substance of the book itself continued to be argued and debated, but only in defense of or opposition to the actual procedure of the General Assembly. One of the earliest publications, *A Letter to a Gentleman at Edinburgh,* focused almost entirely upon the procedure. Starting with a statement of disbelief that this was "the very same Book [that] has for nigh these Eighty Years, been suffer'd, without Let or Molestation, but from the *Baxterian* Side, to pass and repass thro' *Britain* and *Ireland,*" the author affirmed that the General Assembly's act must have been a party action. The general body had had no warning of the matter and, essentially, was asked to vote upon something that most members had not read. The author further listed the many eminent divines who had recommended the book throughout its various editions. Perhaps his most important claim was that the church was wasting its time over this dispute. Considering all the errors floating about the contemporary Protestant church—Arianism, deism, popery—the avid censorship of a reasonably tolerable, long accepted book seemed a misdirection of effort.[16]

Advocates for the assembly responded with stern expostulation upon the rights of the church to prohibit publishing and recommending erroneous treatises. Following upon an act of the General Assembly of 1647 and upon basic Presbyterian polity, the church had every right to ask questions and investigate the doctrines of its members in order to guard against heresy. Very little else was offered in defense of the assembly. No one could aver that the original vote was informed in the sense that members had read the book, for very few of the ministers were, in fact, familiar with the substance. Rather the assembly was forced back upon the justice of administrative procedure: those best able to judge the theology had read the book and their recommendations could be trusted.[17]

Just as the evangelicals emphasized the injustice of the assembly's proceedings, the moderates based their arguments upon their disapproval of the *Marrow*'s substance. One of the best statements of the assembly's position was found in *A Friendly Advice for Preserving the Purity of Doctrine and Peace of the Church.* The again anonymous author accused Fisher of confusing moral law, rightfully a part of Christ's law, with the covenant of works. If a distinction between the law of Christ and the moral law were to be maintained, then there would be no means of refuting the antinomian paradox.

[I]n the Covenant of Works, Death was threatened against every sin, without
Hope of Relief from that Covenant: But in the Covenant of Grace, there is a
Relief provided unto Believers, against the Curse threatened in the Law, and
which even their Sins do deserve.

Along more substantial lines, the pamphlet attacked the *Marrow*'s assertion that the essence of faith was assurance. Here *Friendly Advice* was
exceptionally clear: by saving faith was meant belief of those truths proposed in the gospels. One could have saving faith without assurance. "Not
all True Believers have it. . . ."[18]

These formulations made clear the direction in which moderate theology was heading. Growing out of the new emphasis upon moral philosophy encouraged in the universities at both Edinburgh and Glasgow, an
emphasis strengthened by the increasing popularity of Francis Hutchison at
Glasgow, the moderate theology shied away from discussions of free grace
and saving faith. The moderates needed to believe that human beings were
active participants in the world, not the passive recipients of grace. Therefore they preached upon the moral law, encouraging the individual to
refine reason and direct the will in conjunction with the will of God. While
acknowledging that perfection without grace was impossible, these ministers nonetheless believed that a person could partially transform his or her
lifestyle and, through individual effort, begin to approximate a holy life.
They held the notion that good conduct brings the sinner close to God and
enables the sinner to open up to Christ's saving grace, and they justified
their position with the simple statement, generally accepted as true, that no
one who was saved could follow any path but holiness. They came as close
as possible, without actually transgressing Calvinist bounds, to saying that
a person could effect salvation. No other position could be reconciled with
their belief in a loving, reasonable God and their respect for a reasonable,
ordered society.

Considering the speed with which the official theology of the Scottish
church was turning toward the moderate view, the evangelicals were quite
justified in feeling threatened. From their perspective, the moderate party
was betraying the standards by which the Scottish church had stood for
more than a century. Yet the moderates assertively contradicted that accusation. *Friendly Advice* depended heavily upon the Westminster Confession. Unlike the English nonsubscribers, with whom they shared theological affinity, the Scottish moderates invoked both the confession and the
Bible, almost as if the two works held equal authority for believers. Moderates did not lose themselves in arguments over a person's ability to
interpret the Bible; they simply accepted the Westminster Confession as
the canonical interpretation and argued about the interpretation of that.

Curiously, in England the more evangelical party wanted to enact a more structured institutionalization of the nonconformist church, while the progressive members balked at such structure. In Scotland the moderates invoked church structure, while the evangelicals were put in the awkward position of opposing the acts of official committees. Despite this peculiar disparity, the two conflicts had common features. In both cases challenges were waged against the progressives. The progressive clerical party in both Scotland and England had a clear majority over the traditional, and each was therefore able to manipulate the existing institution to ensure its political victory. Nevertheless, for all the rhetoric upon religious freedom, doctrinal purity, and church government, there was more at stake than church politics, even more than sound theology. The progressives were calling for a change in the very core of spirituality, and the traditionalists were standing firm for religious experience as they had always known it.

In 1723 *An Essay Upon Gospel and Legal Preaching* appeared in Edinburgh, in response to an accusation made by several clergy that the General Assembly encouraged "legal" rather than "gospel" preaching. Legal preaching was the accusatory nomenclature that developed as a label for Arminian doctrine. It was recognized by all sides as a negative quality, always set in opposition to gospel preaching. The quality of gospel preaching, of course, changed according to the viewpoint of the writer, for the defining characteristic of gospel preaching was that it was scriptural; it was true. Briefly acknowledging that legal preaching was any contrived "In the Strain of the old Covenant of Works," the author of the *Essay* proceeded to affirm that "we can't be good Christians while we are immoral Men," and that he was not persuaded that legal preaching was

> urging the Observance of God's Law, and the great Duties of Morality, or preaching Faith and Repentance as necessary in order to our Justification . . . and that Believers may be wrought upon by their Hopes and Fears, or preaching the Necessity of Holiness, in order to our Happiness; the Conditionality of the Covenant of Grace, and that a Believer's Sins make him liable to Death.

His discussion of the necessity of repentance in order to obtain forgiveness solidly limited God's mercy by human conceptions of wisdom, holiness, and purity. The author continued by citing Francis Hutchison and others to sustain his exposition upon natural qualifications as opposed to gracious qualifications. The Calvinist paradox was the critical issue, and the rational theologians resolved that human beings, though recipients of God's grace, had to be actors.

> Tho' the grace *to believe* be from God, this does not hinder it from being our *Act* in us, and required of us in order to *Forgiveness*. For if we will not have

it to be a Condition required of us, because the Grace to believe is given of God, then *Holiness* must not be a *Qualification* required of us for our Admission into Heaven, because the Grace to be holy is from God. According to this Logick, nothing can be required of us in order to our Happiness. We must be entirely passive from first to last. *God must work all and we must do nothing,* which is the fundamental Error of the Antinomians.

The moderates had left all discussion of free grace and Christ's redemption in order to concentrate upon the perfection of human virtue. After all, "The Gospel supposes us rational Creatures."[19]

A review of the *Essay* was published anonymously by Ralph Erskine that same year. Erskine challenged the author's definition of legal preaching, claiming that legal preaching was more easily recognized by its omissions. If a preacher "habitually neglected" the essentials of the fall, the corrupt nature of humanity, or union with Christ, the fountain of holiness, he was a legal preacher. Erskine returned to *The Marrow of Modern Divinity* as a supreme example of gospel preaching and devoted much effort (again) to justifying its content.[20] What at first may have seemed a difference in emphasis was actually a basic disagreement over the spiritual path toward salvation. The author of the *Essay* taught that individuals should strive to structure their thoughts and actions in the way of moral virtue, implying that refusal to do so would prevent one's salvation. Erskine believed that individuals should simply open themselves to God's grace, that they must be passive from first to last. To assume that one could voluntarily take a step toward salvation denied humanity's ultimate dependence upon God. Erskine wrote that

> Sincere, tho' weak Christians, can find the sweetness of a true Gospel Sermon, feel Power attending it, and be thereby excited both to the Exercise of Grace and Practice of Holiness, and in like Manner can Discern the Unsavoury Tang of a Legal preachment.[21]

This dependence upon the Holy Spirit's moving within individual souls and the resulting religious emphasis upon emotionally charged piety had dominated Scottish Christianity since the early seventeenth century. By 1720 the public majority of the ministry had chosen to focus their spirituality upon right reason and rational morality.

As might have been expected, considering the diverse backgrounds represented in the community, the Irish Presbyterian church experienced the most severe upheaval of the three British communities. In Ireland the anxiety concerning Arianism merged with a dissatisfaction over the intellectual preoccupation of some ministers and produced a single, critical conflict through which orthodoxy, tradition, and piety became a single force fighting against an opposition perceived variously as heresy, progress, and rationality. All the threads of ideological, institutional, and spiritual disputation were tied together in the largest and longest subscription controversy of the early eighteenth century.

Subscription to the Westminster Confession was not a new concept to the Irish church of 1720. As early as 1672 the Laggan Presbytery, one of the earliest presbyteries formed in Ulster after the Restoration, demanded that all clerical probationers and candidates subscribe the confession.[22] After the General Synod of Ulster was finally established under the protection of William III in 1691, various presbytery minutes indicated that while probationers' subscription was left to the presbytery's discretion, for the presbytery retained its authority concerning ordination decisions, many presbyteries required some formula of subscription. By 1698 the Synod of Ulster had resolved unanimously "that Young Men, when licens'd to preach, be obliged to Subscribe the Confession of Faith, in all the Articles thereof, as the Confession of their Faith." The Synod of Ulster, in 1705, although leaving individual decisions within presbyterial hands, passed a resolution requiring written subscription to the Westminster Confession by all candidates for the ministry.[23] Nevertheless, presbyteries continued to demand or not demand signatures according to their individual traditions.

Also in 1705 John Abernethy, a recent graduate of the University of Glasgow, formed the Belfast Society. A voluntary society organized to promote intellectual inquiry, the Belfast Society boasted a membership comprising the brightest young ministers in northern Ireland. Though not entirely composed of urban ministers, the membership was skewed geographically, with its very proximity to Belfast abetting the development of a strong, united subcommunity inside the larger clerical community of Ulster. Supposedly embarking upon intellectual pursuits for their own sake, the society read and discussed writings including the orthodox, the questionable, and the truly heretical. Yet, as long as discussions remained within the closed circle of the society, there was scarcely any trouble, except perhaps the growing resentment of those ministers, especially those near Belfast, who were excluded.

When the British government under the new rule of George I considered a law granting Irish nonconforming clergy religious liberty, some discontent surfaced. As in England, this liberty would be granted upon subscription to a formula that would guarantee orthodoxy in certain basic areas of Christianity: the Trinity, the deity of Christ, and the sufficiency of the scriptures. Many Presbyterian clerics simply wanted the Westminster Confession to be accepted as the standard, but this promised to be unacceptable to the civil government. This consideration instigated a further debate within the synod about the acceptability of signing any articles of the Church of England, even if those signed were in agreement with the Westminster Confession. The Dublin Presbytery proposed to the synod the following formula that acknowledged the doctrinal tenets in question:

> I Profess Faith in God the Father, and in Jesus Christ, the Eternal Son of God, the true God, and in God the Holy Ghost; and that these Three are One God, the Same in Substance, Equal in Power and Glory. I believe the Holy Scriptures of the Old & New Testament were given by Divine Inspiration, and that they are a Perfect Rule of Christian Faith and Practice; and pursuant to this Belief, I agree to all the Doctrines common to the Protestant Churches at home and abroad, contain'd in their and our Public Confessions of Faith.[24]

After much discussion the synod decided that it was not relinquishing the Westminster Confession by requiring subscription of a formula, especially since the formula was in perfect agreement with the confession and implicitly acknowledged it. Moreover, to subscribe such a formula was lawful if this was demanded by the civil authority as a term of toleration. Yet Francis Hutchison reported in 1718 that

> the pulpits are ringing with them as if there hearers were all absolute Princes going to impose tests and confessions in their several territories, and not a set of people entirely excluded from the smallest hand in the government anywhere, and utterly incapable of bearing any other part in persecution but the sufferers. I have reason however to apprehend that the antipathy to confessions is upon some other ground than a new spirit of charity. Dr. Clarke's book I'm sufficiently informed has made several unfixed in their old principles, if not entirely altered them.[25]

Actual battle was engaged upon two events: the delivery of a sermon and the receipt of a letter. In December 1719 John Abernethy preached a sermon entitled *Religious Obedience founded on Personal Persuasion*.[26] Abernethy stated that theological diversity came from different, unequal experiences acting on imperfect human beings. Affirming the traditional belief in human inability to know the mind of God, he concluded that

> the Decisions of Men are not infallible Declarations of [God's] Mind, and we cannot be safe in submitting to them Absolutely; tho our doing so may secure us against the reproaches of the world, yet it will not be a sufficient Defence against the reproaches of our Consciences or the Displeasure of God.[27]

Thus each individual must use the full powers of reason to understand the way of salvation as presented in the Bible. Recognizing reason as within the capability of every person, Abernethy had to face the long history of theological disagreements among Protestants. His resolution involved a distinction between essential and nonessential matters of faith, "wherein the Reason of Men and the sincerity of Christians permit them to differ." The infidel's "sin does not consist in the Error of his judgment, but in the Attachment of his heart to his lusts and worldly interests, whereby he is misled and prejudiced against the Truth."[28]

With a growing dependence upon reason, Abernethy and his colleagues developed a contradictory skepticism about the ability of human beings to discern the truth at all. If human reason is perceived as the primary tool of theological investigation, and if human reason is recognized as fallible, then beliefs previously considered "true" are suspect, and open disagreement will be a natural, legitimate result. Thus, following this skepticism was a new conviction that Christian communion should not be based upon the conformity of one's beliefs to the mainstream, but upon the sincerity, as judged by the reasonableness, with which a Christian seeks the truth. Excepting public transgressors, "who in any other like instances notoriously transgress the PLAIN and ESSENTIAL precepts of Jesus Christ," for they "*sin against their light*, and are *self-condemned*," Abernethy concluded that as the Apostles "had *no dominion over the faith* of *Christians*," the modern church had no right to assume any greater authority.[29]

In response to anxiety raised by this sermon, the Synod of Ulster passed in 1720 a series of "Pacifick Acts" to reassure the people. The resolution began "whereas there has been a Surmize of a design to lay aside the Westminster Confession of Faith, and our larger and shorter Catechisms—We of this synod do unanimously declare that none of us have or had such a design," in fact, they still recommended it to all communicants as an excellent compendium of Christian doctrine. The synod reaffirmed the act of 1705 requiring all prospective ministers to subscribe the Westminster Confession, "which is thus to be understood as is now practis'd by the Presbyteries." Further, if any candidate had scruples with any phrase in the confession, he was permitted to use his own words, as long as the presbytery judged his words sound. Actually this act caused no tightening of subscription requirements, since it allowed presbyteries to continue doing as they had done. Finally, all members agreed to refrain from publishing controversial papers on this matter, and instead to work toward unity and peace.[30]

The second event was directly linked to the convocation at Salters' Hall. One congregation of Belfast had called Samuel Haliday as pastor. Since Haliday's appointment was agreeable to James Kirkpatrick, the minister serving the neighboring congregation, Belfast looked forward to a new friendship between the two congregations. After the call had been sent by the Presbytery of Belfast and accepted by Haliday, the presbytery received a copy of a letter from Samuel Dunlop, minister at Althone, to one of its members, accusing Haliday, an observer at the Salters' Hall conference, of Arianism. The matter was under presbyterial consideration throughout the first half of 1720, until Haliday himself brought the accusations to the General Synod and demanded justice.

In response to an inquiry from Kirkpatrick, Dunlop had affirmed that "Mr. Haliday joyn'd the Arian Party," by which he meant "the Non-Subscribers in London, whom he alleages to be generally suspected of Arianism," and further, that "Mr. Haliday's conduct was resented by many pious and orthodox Ministers in London. . . ." Haliday had responded that he had joined neither side at Salters' Hall since he was a stranger to the place (he was merely traveling from his previous pastorate in Rotterdam to his new charge in Ireland), and he declared himself against Arianism, as he had done before. In synod, Haliday now produced letters vindicating his conduct and orthodoxy by widely respected divines, such as Edward Calamy, and a laborious questioning of Dunlop demonstrated a paucity of evidence against Haliday. The final blow was struck when Colonel Upton, probably the elder with the greatest authority and influence, spoke in favor of Haliday's conduct in London. The synod vindicated Haliday, rebuked Dunlop for "rash and impudent behavior," and decided to publicly announce its decision to the Belfast congregation.[31]

The following year saw a steep escalation of contention. In a letter to Principal Stirling of Glasgow, Robert McBride noted that five ministers in the Belfast Presbytery had protested the installation of Samuel Haliday.[32] Alexander McCracken wrote Stirling that from the coming of Haliday "riseth a jealousy amongst the people that there are some amongst us that are unsound and are falling in with the non-subscribers in England and our confession is now to be laid aside. . . ."[33] Joseph Boyse of Dublin, perhaps the most respected cleric in Ireland, recounted to Stirling the bitterness of the struggle. Sympathizing with both points of view, he nonetheless condemned the continuation of strife, especially noting Alexander McCracken's paper, which he found "very uncharitable . . . tending to increase rather than extinguish the flame of contention," and asserting "several things utterly untrue."[34] The orthodox Robert Craighead had the previous June preached at Belfast *A Plea for Peace*, in which he sympathetically formulated the two viewpoints as, first, "without this [subscription] Error and Heresy cannot be sufficiently guarded against," and on the other side, "no form of humane Composure should be set up as an invariable Standard of Orthodoxy."[35] His attempts to balance the demand for Christian liberty with the need for integrity of doctrine were weighed against the causes and consequences of violent contention. From an overly zealous protection of either truth or liberty, charity and reason would be lost in passion, pride, and malice. Divisions occurred when personal attacks and the use of institutional discipline took the place of reason and charity. The scandal and reproach that would result would lead to corruption and subversion of the church, provoking God to anger. Yet how could any such argument have influenced a man like Alexander McCracken, who sincerely believed that "our Confession is struck at in doctrine, worship, discipline, and government. . . ."[36]

Nor was discontent limited to individual clergymen. Whether scandalized by the theology preached by the members of the Belfast Society, enraged by the publication of polemical pamphlets, or inspired by contentious preachers such as McCracken, the laity quickly and decisively entered the fray. George Lang told Robert Wodrow that he feared that those against subscription were sympathetic to Arian views. He had heard some of them argue for Arianism, presumably for the sake of argument, and a few had said that they would not refuse to communicate with Arians. He had concluded that "however it be as to Arianism . . . several ministers incline to Arminian principles." Yet no matter how seriously Lang questioned the Belfast Society's orthodoxy, his most earnest concern was its effect upon the laity: not corrupting, but divisive.

> The manner of the Society's propagating their principles in public sermons and otherwise hath so inflamed the generality of the people and made them so jealous of ministers that those who incline to moderate and peaceful methods have a very difficult part to act. . . .[37]

The Synod of 1721 responded to the fervor with discussion, a few decisions, and several important nondecisions. The first challenge thrown down was an overture concerning the deity of Christ. Introduced with the words "Whereas several aspersions have been cast upon the Protestant Dissenters of our Communion in this Kingdom," the approved overture affirmed the synod's belief in "the Essential Deity of the Son of God," "His Essential Divine perfections, particularly His necessary Existence, Absolute Eternity & Independence" and declared that any person proven to have denied said article would be cast out of the membership. A significant minority refused to debate this resolution, explaining that while they believed strongly in Christ's deity, they were not willing to join in this resolution, because they "are against all Authoritative Decisions of Human tests of Orthodoxy, & because they believe such decisions to be unseasonable at this time."[38] The second problem brought before the synod was the protest of a portion of the Belfast Presbytery against the installation of Samuel Haliday. All the favorable evidence from the previous year was again presented, along with letters from respected men and presbyteries throughout Ireland and with a certificate signed by a "great number" of Belfast congregants attesting to Haliday's orthodoxy. The long debate was suddenly interrupted with a motion that Haliday be asked whether he still adhered to the assent he had given to the Westminster Confession when he was licensed. He refused to commit himself, not

> from my disbelief of the important truths contained in it, the contrary of which I have oft by word & writing declared . . . But my Scruples are against Submitting to Human Tests of Divine truths, Especially in great number of Extra essential points without the knowledge & belief of w^c men may be entitled to the favour of God . . . when Imposed as a necessary term of such Communion.

The synod righteously proclaimed its readiness to hear Haliday's reasons, affirming a dread of "Sinful imposition," when the commissioners from Dublin recommended that the entire matter be dropped. So it was, with the further declaration that however irregular the installation was, as Haliday had explicitly submitted himself to the rule of the synod he would be received as a member.[39]

The most disruptive action was initiated by several orthodox members: a proposal that all synod members willing to subscribe the Westminster Confession be permitted to do so within that public forum. Significantly, this motion was not made until Saturday, and despite the protest of several Belfast ministers and the peacemaking Dublin commissioners that more time for discussion was required, the majority was anxious to vote. Many had planned to leave the synod that afternoon in order to return to their pastoral duties on Sunday, and they wanted an opportunity to publicly subscribe the confession. The nonsubscribing brethren hoped to delay the decision until Monday, since most of the members would have returned to their congregations and the nonsubscribing view might have succeeded. The subscribers pushed for a vote, won the issue, and voted to allow a voluntary subscription.[40]

The following Tuesday reconciliation was attempted with the passage of two charitable declarations. The first, prepared by Joseph Boyse and endorsed by all who protested when Saturday's session called for a vote, stated that the signers believed that those who had acted in opposition to them were guided by their consciences and, further, that they would maintain communion with them. In response, the synod passed an overture declaring that the decision to allow members to subscribe the confession was meant to cast no aspersions upon those who could not in good conscience follow suit. The synod did not question their soundness and had no desire to break communion; further, they recommended that the people put aside all doubts on this account.[41] The very next day, as if to prove the futility of such measures, several members of Haliday's congregation petitioned to be separated into a third Belfast congregation. Although the synod at this time recommended extensive councils for reconciliation, it would eventually sanction the new formation.[42]

From here on the pamphleteers knew no limitations. Beginning with Victor Ferguson's *Vindication of the Presbyterian Ministers in the North of Ireland: Subscribers and Non-Subscribers*, the nonsubscribers and their opponents published over sixty polemical pieces.[43] Repeating their views over and over, the authors demonstrated no change in their opinions over the course of the seven years that the dispute lasted, except perhaps increased obstinacy on both sides. Thus a brief summary of the best of the pieces will provide an adequate log of the points argued by each side.

Published early in the dispute, Ferguson's *Vindication* (probably written by Kirkpatrick; Victor Ferguson was only the publisher of the work) was a level-headed, thorough exposition of the religious system espoused by the nonsubscribers, logically consistent throughout. The case began with the assertion that the Holy Scriptures were the Word of God and a perfect rule of faith, and with an acknowledgment of the Westminster Confession as "a most valuable treasure of Gospel-Truth." Immediately following, the author affirmed that Protestant churches always disclaimed infallibility, that every Christian had the right to decide right and wrong individually, and that only Christ could determine church laws and terms of communion. Historically Protestants had opposed Roman Catholics and nonconformists had opposed Anglicans over this very issue of forced agreement to creeds. Would not subscription to biblical truths in scriptural words be the best prevention against error?

After all, creeds could become idols for weaker Christians. With several pages given to the unwarranted involvement of a basically ignorant laity, the general public was portrayed as

> calling upon their Ministers to swear them into the Confession of Faith and quarrelling with them for not doing it, and yet upon Examination of their Knowledge and Sentiments, daily be[t]ray[ing] gross Ignorance of what is Contain'd in the Book they wou'd swear to.

The author ended his discussion with a review of the proceedings of the 1721 Synod, stating that the nonsubscribers fought subscription as an unbiblical imposition and that there was no proof that they were unsound. "Non-Subscription to the Words of any human Confession whatsoever can never prove Men unsound; because subscription to them is nowhere Commanded in the law of Christ. . . ." The proper, scriptural way to issue such scandal would be through fair trial with at least two witnesses. From start to finish the argument was grounded in the incontestable supremacy and sufficiency of the scriptures, but in this reliance was a firm trust, not shared by their opponents, in human reason.[44]

Charles Masterton, the first pastor of the offshoot, conservative Belfast congregation, published in 1725 one of the best-argued outlines of the prosubscription position. Beginning at precisely the same point as the *Vindication*, with the ultimate truth of scriptures, Masterton elaborated upon the concept of scripture truths. The precepts of the gospel, while all true, were subordinated to one another, and the entire system of essential and unessential truths was related. Moreover, since all scripture truths, whether fundamental or not, tended toward godliness, not one could reasonably be considered a matter of indifference. This discussion of the hierarchy and interrelation of scripture truths was fairly well controlled compared to the vitriolic attack upon the nonsubscribers' emphasis upon sincerity. The knowledge of the truth was "an effectual and solid Perswasion of the Truth as it is in Jesus produced in the Mind of the Children of God by the Supernatural Teaching of the Holy Ghost." The truth was real, not an opinion, and "there is not in all Scripture . . . an Allowance to Act upon a False Principle." What was Christian freedom but freedom from the dominion of sin, of Satan, of the world, freedom from threat of the law? This liberty included freedom from human domination of conscience only if human dictates opposed the law of God. The procedures of the church were necessary aids to the development of human reason. "Freedom consists in the Mind taking Complacency in its own Choice of what is good, under the Direction of right Reason; for where there is no Understanding there can be no Liberty." Although neither the civil magistrate nor the synod had the right to interfere with a person's conscience, pastors of reformed churches, in order to serve the needs of an ignorant, needy populace, had a warrant to unite to keep out errors and promote godliness.[45]

The references in both arguments to the commonality of believers was not coincidental. At the outset the masses showed themselves greatly concerned with the progress of the debate. In his *Seasonable Advice to the Protestant Dissenters in the North of Ireland*, John Abernethy attributed much of the official infighting to popular participation: "The popular Zeal (which is very easily inflamed when any Innovations in Religion are apprehended) was wrought up. . . ." The synod, he implied, was embarrassed by such vulgar suspicions. "The poor abused People, like Men in a Pannick, fancy'd they saw whatever they were afraid of, tho' the danger was all Visionary, and a meer Creature of their Imaginations."[46] And no wonder the behavior of the laity was so thoroughly discussed and condemned. Reports from both sides indicated a heavy influence for a group supposedly subservient to the leadership of its better-educated, better-equipped pastor. Lang again wrote Wodrow in November 1721, noting that among the residents of Down and Antrim was great uneasiness; several persons

> do not hear and communicate with their pastors who are of the society; and I am lately informed that they resent the subscribers sitting in the same Presbytery with these gentlemen, the generality of whom do notwithstanding keep their old way, except two or three who have been obliged by their congregations to subscribe and quit the society, which they chose rather than lose their congregations.[47]

The Synod of 1722 was a quiet affair. Five overtures tending toward peace were passed, including an agreement that declarations of faith in scripture words only were insufficient tests of orthodoxy; an adherence to the Westminster Confession as founded upon the Word of God; a reaffirmation of the scriptural authority for Presbyterian government; a forbearance of the nonsubscribers; and a recommendation that people cleave to their pastors. Meanwhile the synod did allow the conservative faction in Belfast to form a separate congregation. Incidentally, the pamphlets kept coming, with spokesmen for the subscribers, such as Samuel Hemphill and Matthew Clerk, plunging in to attack the *Vindication*.[48]

With the increased heat of the clerical combat and the growing uneasiness of the laity, the Synod of 1723 watched as the conflict slipped into a level of personal invective generally not observed within the institutional forum. The new, third congregation of Belfast complained that some members of the other two Belfast churches were refused dismissals by their sessions and thus were unable to join the new church. The two primary congregations were also denounced for not giving up to the separating communicants their portion of cloaks, palls, and utensils, and for publicly branding the separating members as "Schismaticks." Both groups came before the synod as appellants from lower judicature (Presbytery and Synod of Belfast) decisions. The entire matter was temporarily circumvented through the technical ruling that these matters did not come "regularly," that is, through proper procedures, before the General Synod.[49]

The major portion of the General Synod was occupied with another

appeal from a sentence of the Belfast Synod. The two original congregations complained that in a personal application to Scotland, commissioners
from the new congregation were reported to have accused the ministers
currently in Belfast of maintaining principles that "open a door to lett in all
Errors and Heresys." Colonel Arthur Upton not only admitted making
such a statement, he asserted that he was prepared to prove it and that he
had refused communion with nonsubscribers. Before the Belfast Synod he
openly accused the nonsubscribers of heresy. Upton's attack had originally
focused upon Ferguson's *Vindication*, thus creating several problems for
the judicatory process. Could an anonymous pamphlet be used as evidence
against anyone? Against whom was the charge laid? Since much of the
hearing was in the form of an appeal, could new evidence be introduced?
Did the Belfast Synod, from whom appeal was made, have the right to sit
in judgment? In other words, an amazing array of technical questions was
available in lieu of the true subject for debate, and points were either won
or compromised through the technicality votes.

The Synod of Belfast had decided the original trial with three resolutions: first, that only Kirkpatrick, Haliday, Michael Bruce, and Thomas
Nevin, the original defenders of the *Vindication*, were defendants; second,
that the four were not guilty of opening a door to heresy; and third, that
Upton was guilty of breach of charity. The General Synod upheld Upton's
right to appeal all three decisions and then refused Kirkpatrick's request to
enter his protest to the vote. Next, it was voted that the Synod of Belfast
had no right to a voice in deciding the appeal, and again its protest was not
recorded. There then followed debate regarding the appropriateness of
permitting several ministers and elders to sit in judgment, since they had
already, the defendants argued, decided against the nonsubscribers. The
General Synod found in favor of the defendants only once: Matthew Clerk,
writer of two pamphlets against the nonsubscribers, was not permitted to
serve as judge. From this point followed several detailed arguments on
evidence and procedure. In each case the decision went against the nonsubscribers, a predictable conclusion, since the majority of nonsubscriber
sympathizers, a minority during the best of times, were members of the
temporarily excluded Belfast Synod.

Finally, the defendants discovered a reasonable quibble. As the case
was an appeal, they declared that Upton had no right to bring new evidence
without a ten-day notice. This was an established point of procedure, and,
as Upton wanted to present all the old and new evidence together, he was
forced to cease his prosecution. Even without a judicial victory, however,
Upton did experience a moral victory in the four resolutions with which the
assembly ended the affair. The Ulster Synod decided unanimously that

condemning all declarations in human words as tests of orthodoxy did, indeed, open a door to heresy. Second, it was judged dangerous to allow a presbytery to judge a candidate's own formulations in lieu of the accepted standard. Moreover, the *Vindication* was adduced to contain dangerous principles, and its authors, whoever they may have been, were judged disturbers of the peace. Finally, the actual trial was postponed a year, according to the rules of discipline. Even this last decision went against the wishes of the nonsubscribers, since they probably would have been exonerated by an immediate trial based on the old evidence.[50]

The year that followed the Ulster Synod's decision suffered even greater tribulation. In several letters to Robert Wodrow, Charles Masterton recorded the attempts of the nonsubscribers to inflame the dispute. They "use uncommon industry to blacken the conduct of the Synod, and single out some members as the butt of their prejudice, viz. Mr. Upton, Mr. Gilbert Kennedy, myself as moderator [of the 1723 Ulster Synod]."[51] Despite such efforts, the majority of the laity remained on the extreme conservative side. William Livingston told William Macknight that the people had stopped all converse with the nonsubscribers; they would not even hear them preach. When John Henderson was appointed to preach at the ordination of Robert McMaster, Masterton's successor, the people shut the church doors and stopped the ordination: "we had much to do to persuade them to allow the nonsubscribers to lay on hands with us." Vacant congregations refused supply from nonsubscribing clergy and objected against suspicious probationers.[52] While the lay response was railed against in the nonsubscribers' pamphlets, the voice of the people was heard.

The General Synod of 1724 postponed the Upton appeal another year; the major event was the trial of Thomas Nevin, accused of saying that it was no blasphemy to say Christ was not God. The trial occupied several days, during which procedure and verbal formulae were battered about. Nevin argued context and intention while his opponents refused to let go of what he had said. In the midst of the long display, a proposal was approved that Nevin "for the glory of God, the Edification of this Church, and Mr. Nivin's own vindication . . . should make a declareation of his belief in the Supreme Deity of our Lord Jesus Christ." He refused and was deposed and excommunicated, setting off yet another rash of pamphlets.[53]

Clergy and laity continued to throw challenges at the erratically enforced subscription system. Alexander Colville, probationer in the Synod of Armagh, was called to the congregation his father had served, Dromore. He had been licensed by the Presbytery of Cupar in Scotland and therefore must have subscribed, but he was suspected by some of having changed his theological views and was asked to subscribe again. He refused, and his opponents took this as evidence of error. Although the majority of the congregation continued in their call (a clear exception among the laity), the Presbytery of Armagh, supported by the subsynod, refused ordination.

Rather than await trial by the General Synod, Colville traveled to London and obtained ordination and certifications from the liberal London ministers who spoke of the Irish Presbyterian conduct "with the greatest abhorrence as tyranny, persecution, and popery at its greatest height."[54] Colville returned to serve the congregation, and they asked the presbytery to install him, to the dismay of a minority of his congregation and the majority of northern Irish ministers. The 1725 Synod suspended Colville, but not without the dissent of several nonsubscribers who declared that since Colville had never been received as a synod member, he was not under their jurisdiction.[55] Colville placed himself and his supporters in the congregation under the care of the Presbytery of Dublin, which very carefully accepted the charge while trying to act as peacemaker. When the Dublin ministers tried to install Colville and lift his suspension, they met with remonstrances such as Colville's "denying the doctrine of the supreme deity of our blessed Saviour to be fundamental, his speaking prophanely of the doctrine of the Trinity, his neglecting family worship." The Dublin ministers offered to sit in official trial, Armagh refused, and Colville was installed by the Dubliners.[56]

The appeal of Upton was officially dropped, but further overtures were passed to resolve the contention, if not reconcile the adversaries. Subscribers were now to be indulged in their conscientious scruples against communion with nonsubscribers, and, after some consideration, nonsubscribers received the same indulgence. (Notably, seven ministers and six elders, including the cleric Matthew Clerk, dissented from the second half of this overture.) A second overture certified that allowances of the 1720 Pacific Acts extended to phrases only and not doctrines. Any candidate scrupling a doctrine was to be referred to the general meeting; any presbytery moderator disregarding this rule was subject to a year's suspension. The nonsubscribers found this resolution severe, explaining that it contradicted the facts of the Pacific Acts and arguments against subscription, took away power from presbyteries, violated the rights of congregations, and prescribed excessive punishments for infractions. The final resolution, and most ominous, involved reassigning ministers into new presbyteries, with all the nonsubscribers gathered into the new Presbytery of Antrim.

The other sign of impending schism was the allowance of separate interloquiters for subscribers and nonsubscribers. The interloquiter, an informal clerical caucus originally designed to afford opportunities for the clergy to meet without elders and settle matters, such as the regium donum, that did not require lay input, had become a forum for developing resolutions and streamlining problems. The justification for separate meetings was the hope that each group would formulate overtures for reconciliation. The subscribers offered five overtures, all of which were deemed outrageous by their opponents. First, anyone casting reproaches on the church judicatures was subject to discipline, and members of the clergy were liable to suspension. Anyone not accepting the authority lodged by Christ in the hierarchical judicatory system was denied a vote. Anyone refusing to declare his faith upon any given article of the confession, when such declaration was judged contributing "to the glory of God, and the edification of souls," would be censured. Fourth, the synod should consider following the example of the Church of Scotland and demand subscription. Finally, no judicature was permitted to overturn the decisions of a superior court.[57]

The following year the nonsubscribers brought five resolutions of their own before the General Synod. They first asked that all subscribers consider the nonsubscribers' position, which they then outlined with extensive argument. The terms of communion were set by Christ. Candidates for the clergy should be permitted to use their own words in making statements of their faith, and each presbytery had the duty to ordain qualified candidates with or without their subscription. Neither parents of children to be baptized nor present members should be forced to subscribe. Moreover, voluntary subscription ought to be avoided as dangerous—it worked to cast a "popular Odium" upon nonsubscribers. The problem was the manner in which this was presented to the assembly; many subscribers got no farther than the six-point exposition of nonsubscribing principles before they refused to consider the paper. The other four items, including the ceasing of public debates in presbytery and synod, the end of the pamphlet war, the exercise of charity, and prayer for peace, all disappeared under the horrors of that open statement in direct opposition to the Pacific Acts. Without much debate the synod voted to separate.[58]

An interesting comment on this affair was the responsive behavior of the Southern Association. The presbyteries of Dublin and Munster both continued communion with the ousted Antrim Presbytery, condemning the conduct of the Synod of Ulster. The Dublin commissioners had acted the role of peacemaker, in 1721 urging that the case of Haliday be dropped, in 1723 arguing for postponement of the Upton appeal. By 1725 they had become partisans, sponsoring Alexander Colville at Dromore and working with Edward Calamy and others in London to threaten removal of the royal grant if reconciliation was not achieved. This partisanship was especially distressing to the subscribers, since Joseph Boyse of Dublin was respected as the leading minister in Ireland. He had proven his righteousness in his handling of the Emlyn trial, yet he was now championing suspected heretics. Robert McBride, in *The Overtures transmitted by the General Synod of 1725 Set in a Fair Light*, had to acknowledge a letter by Boyse in support of the nonsubscribers. McBride actually wrote that he assumed Haliday had used Boyse's letter without his knowledge, since Boyse would never have consented to such use. In response, Boyse published *A Vindication of a Private Letter* in which he not only attacked McBride for commenting in print upon private correspondence, but also declared the letter to be his. The subscribers just could not understand the defection of a tried and true man of God.[59]

It is very difficult to comprehend either the progress of events or the true nature of the controversy. The argument put forward by the nonsubscribers was carefully grounded in scripture and logically structured. All authority of religion was based in personal conscience, which the church had no legislative authority to bind. Imposition of creeds was unwarranted by the gospel and unnecessary, since the truth was divinely inscribed in the heart. The nonsubscribers always insisted that there were no doctrinal differences between them and the others, and if any were suspected of heresy, they should be tried presbyterially. More specifically, the subscribers were accused of innovation and schismatic tendencies, of denying reformation principles. If the Christian church had always guarded liberty of conscience, the nonsubscribers argued, the Roman Catholic church would have remained free from corruption. Subscription did not keep the Scottish church free from error, as evidenced by the creeping in of antinomianism. Moreover, the nonsubscribers did in the end speak against the Westminster Confession as too detailed and abstruse.

Had the nonsubscribers merely developed an irrational fear for their religious liberty, or was their freedom in danger? And if these preachers were in danger of censure, would excommunication, deposition, or even admonition have been justifiable as preclusion of heterodoxy? Clearly their opponents thought so. They had answered the proliberty arguments with statements emphasizing the church's duty to maintain purity of doctrine. They believed that interpretation of the Bible should be standardized in order to withstand human frailty and false prophets, and they claimed to accept the entirety of the Westminster Confession as that interpretation. The rights of the church superseded those of any individual cleric or lay person. Yet amidst such technical response to the nonsubscribers ran a certain thread of suspicion. They would not state publicly their beliefs; they would not clarify their terms. They found some articles unessential and would not say which ones or justify their conclusions. If the non-subscribers were orthodox, why did they adamantly refuse to reassure their colleagues? How would they be compromising themselves?[60]

As in England and Scotland, the progressive theologians had placed themselves against the conservatives. A great number of moderates, seeking peace and stability, including initially the Southern Association, watched while the exclusive Belfast Society fenced with vehement conservatives like Clerk and Masterton. The issue was again brought before the church courts, and the progressives again chose to fight through institutional structures. In England, the progressives fought subscription as the addition of structure to a traditionally unstructured communion; they won. The Scottish progressives stood for conformity in an established church that had always sanctioned institutional forms; they also won. In Ulster, however, the progressives had used the logic of English divines to combat a Scottish-style institution. They may have claimed that this rigid organization of the Ulster Synod was a novelty of the eighteenth century, but in fact the majority of Ulster Presbyterians had always understood that the church was an official body with rights, privileges, and procedures independent of individuals. Rather than play according to the rules, or even attempt to change the rules, the nonsubscribers acted to convince their opponents and the undecided that the order of things had been completely misunderstood all those years. They lost; tradition could not be quite so openly flouted.

As in England and Scotland, progressives and conservatives were fighting for dominance of the church. In Ireland, the nonsubscribing party was the home of the intellectual elite. Joseph Boyse, John Abernethy, Michael Bruce, James Kirkpatrick, for example, all not only published extensively during the controversy, but produced treatises and sermons for the press afterwards. Abernethy was the author of a four-volume systematic theologia. Most of the pamphlets published by the nonsubscribers were well documented by scripture and the writings of church fathers, precisely and impeccably structured, and elegantly written. In comparison, the subscribers' pamphlets were confused, awkward, and childish. Against the lucid and calm erudition of nonsubscribing essays, the subscribers, with the exception of Masterton, read like angry children spouting indiscriminately. The high quality of the nonsubscribers' thought should have been matched with political insight. Why did they act in ways guaranteed to fail?

The nonsubscribers were trained outside Ireland. Those who came from England, including the southerners, were raised within traditions that had opposed the church establishment in the guise of the Church of England. Creeds and subscriptions were tools of the traditional enemy, an enemy also present in the Church of Ireland; the battle cry was always shouted against the Anglican church as one step removed from Rome. The overwhelming majority of Irish ministers, however, were trained in Scotland, especially Glasgow. Here the church was established; pro forma subscription was established. Yet most of the ministers, certainly the professors of divinity, were of the moderate persuasion. Thus Irish candidates were trained within a progressive theological system that was supported by the church structure. It is no coincidence that the best-educated constituted the nonsubscribing party. Following the example set in the Scottish church, the progressives had hoped to accomplish their goals through institutional channels. The Pacific Acts of 1720 indicated that this might have been possible. Only after the judicature turned against them did the nonsubscribers begin to argue within the bounds set at Salters' Hall.

Since most of the clergy were trained in Scotland, the Irish community could have been expected to be a Scottish miniature. How was it different? The answer lies in the narrative itself. In 1720 Samuel Haliday was vindicated with ease and speed; his slanderers were rebuked for imprudence. That same synod saw the passage of the Pacific Acts, a series of five resolutions designed ambiguously in order to reinforce the status quo. The very next year Haliday's installation was challenged, the synod passed a resolution affirming the divinity of Christ, and a vast collection of clerics asked permission to voluntarily subscribe the Westminster Confession. This contradictory legislation was due to the movement of the laity. Some congregants in Belfast were anxious because Haliday had not resubscribed after his—to them—questionable behavior at Salters' Hall, even though his conduct had been vindicated the previous year. Both the deity of Christ and the voluntary subscription resolutions were passed in response to pressure from congregations that the church declare itself on these crucial issues; as early as 1721 ministers were reporting such pressure to their colleagues in Scotland. Colonel Upton, an elder, brought the charges against the four nonsubscribers, and the laity sought a third congregation in Belfast, one sympathetic to their views. Reports to presbyteries and synods said that parishioners refused to hear nonsubscribers' preaching, and refused communion with them; General Synod resolutions from 1722 made allowances for individuals who scrupled to pray with nonsubscribers, allowances in direct contradiction to Presbyterian discipline. Even the unfortunate Thomas Nevin had been accused originally by a lay person. When the synod in 1726 voted on the question of separation, the minutes recorded that a large majority favored division. The clergy voted 36–34 in favor, two ministers not present recorded their protest the following Monday, rendering the vote a tie, and 42 ministers abstained. The large majority must have been composed entirely of elders.[61]

Of what were the parishioners anxious? The obvious answer is Arianism, yet there is really no evidence in the copious writings of the nonsubscribers to suggest that they followed in that direction. The laity used terms such as "Arian," but it is unlikely that the average lay person possessed the ability and education required to distinguish one heresy from another. A better explanation might be that "Arian" became a code word for heretic, a generic label for anyone who did not preach along traditional lines. Like the moderates of Scotland, Irish ministers in general were moving in Arminian directions, including subscribers as well as nonsubscribers. Most were preaching on moral virtue, paths of holiness, and the potential of humanity to satisfy an amiable, reasonable God. Part of the nonsubscribers' argument emphasized the virtue of sincerity over truth. As they laid out the position, they could not believe that

> . . . an Infinitely Wise, Just, Holy and Gracious God may Eternally Punish wise and good Men ACTING SINCERELY according to their Conscience, and the best of their Capacities and Opportunities merely for an Involuntary Error in Judgment *Consistent* with Christian Sincerity and true Piety; But this is very Absurd & tends to promote very unworthy thoughts of the Supreme Being, and of his moral Perfections and Government.[62]

The nonsubscribers' God was a different deity from Knox's awesome, omnipotent, arbitrary God; their God was knowable, likeable, eminently reasonable. The hearers of such words were not comforted. As one Irish parishioner said of Francis Hutchison Junior to his father:

> Your silly son, Frank, has fashed a' the congregation with his idle cackle, for he has been babbling' this 'our about a good and benevolent God and that the souls of the heathen themselves will gang tae heaven if they follow the licht o' their aen consciences. Not a word does the daft boy ken, spur, nor say about the gude auld comfortable doctrine of election, reprobation, original sin, and faith.[63]

During this era of progress that threatened the core of religious experience, subscription was a visible handhold on truth. The Westminster Confession was part of the Scots-Irish tradition. The confession was the recognized digest of orthodox theology, not because the believers had read it and found it good—those who *were* studying it found it wanting—but because Robert Blair, John Livingstone, Robert Baillie, the fathers of the 1638 reformation, had found it good. The document itself had come to be a traditional symbol of right religion. No matter how peculiar the dogma a minister preached, his hearers were certain that if he subscribed the Westminster Confession, he was orthodox, traditional, a practitioner of the "Good Old Way." Moreover, unlike the Church of Scotland, where subscription was pro forma and thus almost insignificant, the Irish Presbyterians, through public debate, had elevated subscription to a meaningful act. The refusal of the nonsubscribers to sign enabled others, perhaps just as progressive theologically, to demonstrate easily their fitness. Many ministers sympathetic to the nonsubscription cause admitted that they had subscribed only in response to congregational pressure. Nevertheless, the reasons were not known and thus not as important as the act itself. For a time Presbyterians throughout Ireland were satisfied as long as the people believed in the righteousness of their clergy, and the symbol of subscription stood. When parishioners discovered subscription to be a false, empty symbol, this unity dissolved. The Presbyterian churches in Scotland and Ireland would both suffer as the forces of progressive, rational theology opposed those of traditional, emotional piety without the mediation of a symbolic orthodoxy.

4

Piety Against Reason

In April 1720 Hugh Cumin, a lay member and elder of the congregation of Cambuslang, presented a protest to the session of his church. He began with the almost generically correct mode by complaining of defections from "our covenant engagement, national and sollemn league and covenant." He objected to the union with England as "a way to introduce us to their ceremonies" and to the abjuration oath as, again, "contrary to our covenant, w^c were once the Glory of our church and land w^c we now regrate is much forgot." Scandalous persons were elected to church offices and admitted to participation in the sacraments; elders ruled privately rather than through an open, public forum. Most telling, Cumin accused ministers of "in baptism, injoining parents to train up their children in the protestant religion not making a difference betwixt presbyterian protestants and episcopal protestants."[1]

Also in the west of Scotland, a young preacher named John Adamson, originally of Perth, had broken from the communion and was touring the countryside enjoying popularity as an intinerant. As he was far beyond the bounds of his license, it was taken away, and, eventually, he was excommunicated. In 1723 Adamson appeared again in central Scotland, in the Auchterarder Presbytery. He had set himself up in Fossway, successfully attracting people from other parishes to his organization. Individual pastors warned their people against hearing him: he was a schismatic, a usurper of ministry, immoral, and cast out and delivered to the devil upon his obstinacy against authority. Anyone attending his sermons was liable to censure. Still people went; several signed up. These lay persons were summoned to the presbytery meeting, and they refused to attend. Commissioners went after them, and the recalcitrants refused to speak to the commissioners.[2]

Hugh Cumin and John Adamson were only two among many who

feared and resented the changes coming to the Scottish church. The moderate party, with a clear majority as early as 1715, was facing an opposition increasingly discontented and anxious for redress. The early battles surrounding *The Marrow of Modern Divinity* were only the more public displays of the competition between the progressive intellectuals with their moral philosophy and the traditional preachers with their saving grace. The fierceness of the combatants and the frequency with which this same battle informed all other topics of church debate give some hint of the import of these discussions to men's formulations upon their own and the community's identity. Each clergyman was fighting for the integrity of his own religiosity and for control over the church and his congregation.

The eighteenth century, an era of toleration and enlightenment, saw these conflicts escalate to fever pitch. The Presbyterians in both Scotland and Ireland would be forced to grapple with such opposite positions, and the resolutions would never be resolutions but permanent division. Ireland in 1726 had already had one "agree to disagree" decision, but more were to come. Adherents to both sides had dug their heels in too deeply to allow compromise or reconciliation. Waging war on the synod floor as well as from the pulpit, ministers fought for their congregations' support. Nevertheless, the laity seemed poorly impressed with the often esoteric nature of many of their arguments. The demands of the people were simple and traditional: a true, reformed, Protestant church that was loyal to and consistent with the ideas and practices of Scottish Presbyterianism, as it had always been since the days of Knox.

The ministers knew of this need; they invoked it themselves as a rhetorical charm. Yet they could not understand of what such a tradition was composed. The national Church of Scotland was less than two hundred years old. It had been Knoxian Presbyterian for thirty years, modified Presbyterian for twenty, episcopalian for twenty-eight, radical Presbyterian for twelve, split in two Presbyterian camps for ten, prelatic again for twenty-seven, and, at last in 1690, Presbyterian forever (though actually by 1720 forever had been only thirty years). The learned members of the clerical profession knew that most people had been content during the early years of the first prelatic period and that the Presbyterian ministers who were so successful during the early years of the Ulster plantation were ordained within the prelatic Church of Ireland. The clergy also knew that throughout the tribulations of the seventeenth century most people simply ignored the differing religious opinions and got on with the business of the church. That the fight against episcopacy during the Restoration had been strong no one denied, but they also could not deny that many clergymen who had conformed in one direction conformed again in the other with little protest. The ministers of the Enlightenment felt free to pursue truth and salvation through their reason; they used their intellects, and they knew that the legend of a single reformed tradition from John Knox to *The Marrow of Modern Divinity* had few bases in fact.

Legends do not need factual bases; they live independently, providing by their very existence their own proof. The people of Scotland believed that one continuous tradition of Protestantism, initiated by Knox himself, had flourished uninterrupted from 1560 to the contemporary period. Sometimes this tradition was predominant, as in the 1638 reformation; occasionally the true religion had asserted itself through an underground movement, such as the conventicles of the 1660s and 1670s. Now in the 1720s and 1730s the people and their clerical advocates had begun to search out that tradition in eighteenth-century Scottish Presbyterianism. The quality of that search was confrontational and sensationalist. Accusations and counterclaims were hurled across the territory with a fury seen only in ideological struggles. The laity occasionally participated, though they were far more likely to be active observers who made their preferences known by affiliation with one side. The context of the battles spanning the years between 1725 and 1760 was the familiar formula that pitted reason against piety. However, the quality of these fights, the problems brought forward, the method of argument, and the sacred pasts invoked all came together to provide the best account thus far of the character of this Knoxian reformed tradition so elusive to definition but so important to the identity of this communion.

By 1725, the Church of Scotland had yet to appease the orthodox Calvinists who feared that the church was falling into the wicked hands of Enlightenment men. The ousting of the nonsubscribers in Ireland had brought a feeling of finality; it proved a watershed for overwrought zeal and popular emotion. Conservatives, both clerical and lay, had been satisfied with that outcome. In Scotland, however, the liberals were still full synod members; in fact, every church decision indicated that these liberals, if not progressives, were ruling the General Assembly. In 1713 John Simson had been tried for heresy; the General Assembly had essentially cleared the Glasgow professor of all charges in the face of strong evidence against him, warning the more staunchly Calvinist that the content of heresy and orthodoxy was undergoing serious alteration. Then the General Assembly thoroughly censured the Presbytery of Auchterarder only for doing its duty in trying to protect itself against Arminian laxity. And the sweeping actions of the assembly's Committee on the Purity of Doctrine had quickly and unceremoniously not only dismissed but condemned a book that had for at least sixty years been accepted by almost all respectable Calvinists.

As far as the evangelical party was concerned, they were the righteous minority. Just because a committee had declared *The Marrow of Modern Divinity* unsound did not mean that they had to accept that judgment. For the next few years, the evangelical party would continue to try to bring the majority to some sense of their errors until, in despair, the most extreme segment chose to leave, or secede, from the national Church of Scotland. The Seceders would invoke the old formulas and traditions, eventually sending missionaries to serve some Irish congregations grown suspicious of even the subscribers' orthodoxy. The Seceders did remain a peripheral, extremist organization, devoting their energy to increasingly outdated causes such as the renewal of the Solemn League and Covenant. Nevertheless, it is noteworthy that the Seceders did attract a vast following, with numerous lay societies petitioning to join their network and enjoy their preaching. If their clerical colleagues were not impressed by the return to traditional, revivalist rhetoric and practices, many among the laity were anxious to join.

The first indication of continued evangelical disaffection was the retrial of John Simson. Retrial is, perhaps, not the correct word, for he was brought forward in 1727 on new evidence of heterodoxy. There was no doubt, even in the minds of the most progressive theologians, that Simson had crossed the boundary of acceptable speculation. The entire process was an embarrassment to the moderate party, for they found themselves the objects of the self-righteous gloating of a group of ministers who could not allow such an opportunity to pass. The process lasted almost three years, involving the hearing of two separate libels. At the beginning of 1729 the assembly's commission asked the presbyteries to send their individual opinions before they recommended a procedure to the General Assembly. Of the two libels pending, only some points of the first had been proven. The assembly had only suspended Simson until all of the charges could be heard. Simson refused to clear himself to the appointed commission, so presbyterial advice was sought.

The Presbytery of Auchterarder voted for deposition, not a continuance of his suspension, from the professoriate and the ministry. He held dangerous doctrines regarding the Trinity and the divinity of Christ, and he had adopted un-Christian methods in defending himself. Considering the beliefs that he had been proven to espouse, the presbytery further challenged the sincerity of that orthodox declaration he had recently given to demonstrate his innocence.[3] The Presbytery of Hamilton, after grudgingly acknowledging the possibility of Simson's great learning and piety, noted that Simson had not obeyed the General Assembly act of 1717 requiring that he be more circumspect in his behavior. They found "much unsteadiness in His principles with respect to the Important and fundamental Doctrine of the Ever blessed Trinity," and that at a time when this doctrine was being challenged throughout Britain. "They heartily wish that when the General Assembly of this Church discharge opinions to be taught or preached a particular condescendance were allwayes made of the propositions so discharged." He had given great offense to many ministers and elders in refusing to explain his principles. In the end the presbytery judged that he ought to be removed from the university because it was not safe to trust him with the teaching of divinity.[4]

The Presbytery of Lanark advised that the assembly should at once proceed to the second libel and in the meantime "They should take Care of the Preservation of the Purity of Doctrine by laying him aside from his Office of Teaching and Preaching by Deposition, or otherwise as they shall in their great wisdom think fit."[5] A more complete record of presbytery-level discussion was noted in the minutes of the Presbytery of Paisley, which also found Simson deserving of severe censure. They believed it wise for the assembly to "declare it unsafe for this Church that Mr. Simson be continued any longer in the Exercise of teaching, preaching, or of any other Ecclesiastical Function within this Church." Their specific recommendation called for the assembly's proceeding to further censure immediately rather than continuing the suspension while awaiting the outcome of the second libel. They explained that Simson's current suspension was really a necessity considering the charges against him, that it certainly did not fully satisfy as a censure. Moreover, such a declaration as the one they recommended was the only appropriate and adequate response to those parts of the first libel that had been proven. Simson's statements were subversive to the deity of Christ and the Trinity, it was proven that he made these statements, and he had neglected several opportunities to clear himself. Finally, the presbytery believed that such a judgment was required to give clear testimony against such errors. The presbytery called for a formal declaration of the "Divine Glory of Christ."[6]

This case had followed directly upon the problems with one Alexander Scrimsoun, a man who had "intruded" himself as a professor of divinity at the College of St. Andrews. Never ordained or authorized to teach, Scrimsoun had obtained his position through a presentation from the king, granted through the influence of the viscount of Bolen Crook. He had never satisfied any judicature upon the soundness of his faith, and he refused to commune with the presbytery. Although under the sentence of

suspension from teaching, executed by a royal commission of visitation in
1726, Scrimsoun, upon his lawyer's advice, decided to resume teaching
divinity before his libel had been tried.

> [The] royal commission found it of dangerous consequency that a person
> under such strong suspicion of Disaffection should take upon him the
> Instruction of youth training up to the Holy Ministry least he should poison
> them with Corrupt Principles contrary unto the Doctrine of this Church and
> hurtfull to the Civil Government.

The presbytery protested his resuming office and warned that he was liable
to church censure, but such a warning might have little influence upon a
man who had acquired his position through a direct presentation.[7]

The uneasy peace of the church was also threatened from other directions. In 1726 a Commission of the General Assembly was appointed to
"prepare an overture anent the method and strain of preaching introduced
of late by some young preachers very offensive to many of the Lords
people and a Small obstruction to Spiritual edification."[8] In 1730 an act
was passed disallowing the recording of protest until presbyterial consent
had been granted, a measure obviously designed to squelch a very vocal
party in the assembly and a measure that this party would consider an
infringement upon scriptural church government.[9] Although sides were
being chosen in these various conflicts, the major break, the schism, would
result from the problem posed by the Scrimsoun incident—the power of
the civil government over the church, specifically as upheld by the rules of
patronage.

The Presbyterians were angry about many aspects of civil interference. They resented the government's appointment of days of fasting
and thanksgiving and the unbounded toleration granted the episcopal
church. They were outraged that in addressing Parliament they were
acknowledging the authority of English bishops over their affairs, and they
hated kissing the Bible while taking an oath. They disliked the celebration
of festival days, condemned as sinful the balls, games, and fashions that
came with the English nobility, and ardently desired harsher restriction of
the Roman Catholics in the Highlands. Nevertheless, the most onerous
burden these churchmen self-righteously suffered was patronage.

Simply stated, the right of patronage enabled patrons, either landlords
or, in the case of a town, magistrates and council members, to appoint a
pastor. Such an appointment contradicted the clear rights, according to the
Presbyterian persuasion, of congregations to call their own ministers.
While many patrons waived their rights in deference to the Presbyterian
form of government, some, including occasionally Anglicans or even

Roman Catholics, insisted upon exercising their prerogative. The church administration would quietly dispute with such patrons, and frequently the presbytery, synod, or in some cases the General Assembly would have to concede. These concessions were not popular, and they occurred just frequently enough to remind the church members of the problems plaguing their community.

The arguments about patronage were based either in scripture or in Scottish tradition. The scriptural prescriptions for church government established the election of pastors by the congregation, but advocates of patronage argued that the provision allowing for congregational disapproval satisfied that requirement. Oddly enough, the biblical citation stopped at those two points, and the real debate developed around the role that patrons had played historically since the Knoxian reform. All agreed that the 1560 Book of Discipline allowed for the election of a pastor by the kirk, but in 1567 an act of Parliament formally protected the patrons' right of presentation for six months. Both sides also agreed that congregations always had the right of appeal to the judicatures and that the General Assembly's decision was always final; but no one investigated exactly how patronage had operated during these formative, archetypal years.

The acts of 1592 that established legally the Church of Scotland again reserved the right of presentation, but the actual enforcement and use or abuse of this right was unclear more than a century later. Those opposed to patronage invoked the activities of the 1638 reformers. Those supportive undermined the reformers' resolutions because they were passed during the Commonwealth period and therefore not acceptable to any true friend of civil government. In 1690 the law of patronage was reestablished as part of the 1590 code passed down. And so the battle continued over intention and innuendo. (One pamphleteer slyly explained that while popular elections were agreeable to the Word of God, the Church of Scotland had never seen one. If patronage ended, the power of calling a pastor would simply pass to the elders, and where was popular choice there?) Initially the specific issue was a new law of patronage that accompanied the 1707 Act of Union, but after this law was passed in 1712, the problem became the church's implementation of that law.[10]

The Church of Scotland found itself in the peculiar position of an established church, with responsibilities to all classes of people, including the propertied classes, and to the civil government. In order to effect the smooth operation of the law and to decrease the number of conflicts between patrons and congregations, the General Assembly in 1731 proposed an Act Concerning the Method of Planting Vacant Congregations. The act first considered the planting of congregations that fell under the immediate jurisdiction of the presbytery. The presbytery was required to supply parishes with well-qualified men. When heritors or elders requested a call, the congregation was given a ten-day notice before a meeting. Then a moderator from the presbytery would meet with the elders or heritors in the presence of the congregation, and they would call a candidate, subject to the approval of the people. If the congregation disapproved, the matter was referred to the whole presbytery for judgment and could doubtlessly be appealed to the higher judicatures. In the case of royal burghs, election of a pastor was by magistrates, town council, and elders, and, in the case of a landward parish, heritors. All Protestant heritors willing to subscribe to the call (i.e., to commit themselves for the financial support) were given a vote. Again, if the elders disputed the choice of the heritors, the presbytery was to decide.[11]

This act was distributed to the presbyteries for their response, with the addendum that if the majority of the presbyteries approved the measure, it would automatically pass into law. The majority of presbyteries opposed the resolution as restrictive of the church's authority and as a denial of the rights of the congregation. The Presbytery of Paisley, fresh from a dispute over its ordination of a candidate who was opposed by a patron (the presbytery won because of a six-month technicality; having waited seven months the patron had forfeited his right of presentation), objected to patronage in any form as "a Stroak at the Bottom of our Constitution."[12] Auchterarder opposed the measure as it stood but was willing to accept it provided that five conditions were attached. They recommended that elders consult the people and meet by themselves to discuss the candidates and that the presbytery always be consulted regarding the probationer's fitness. They asked that Presbyterian heritors be granted preference over others in the selection process. The rights of presbyteries ought to be protected by allowing them longer than six months in the case of a division or, conversely, permitting the planting of a congregation before six months had elapsed if the patron was willing.[13] Unfortunately, since the presbytery had declared its acceptance of the measure provided those conditions were met, its vote was considered an "aye." Several presbyteries responded with such conditional affirmatives, and the resolution passed into law.

Upon this conclusion four ministers from the Presbytery of Perth and Stirling, Ebenezer Erskine, William Wilson, Alexander Moncrief, and James Fisher, asked that their protest be recorded in the minutes. The General Assembly invoked a previous act of 1730 and refused; the four ministers decried the assembly's conduct and left, suspended for their behavior. *A Publick Testimony*, a pamphlet published anonymously in 1732, probably by these men, provides a brief, biased account of the proceedings. According to the author, when the ministers presented their grievances to the assembly they were referred to committee. Knowing that their complaints would be buried there, the protesters asked to deliver their statement on the floor and were answered that a decision to refer a matter to committee was not reversible. The protesters were about to take up instruments (publicly record their protest) when the moderator raised his hands to pray. So they brought their case to the public.

The pamphlet provided an early outline of the complaints that the evangelical party had against the church, as well as the fervor with which these grievances had been nursed. The authors suspected a conspiracy to reintroduce prelacy in name; it already existed in practice. The Scottish people were victims of "Ecclesiastick Tyranny"; they were deprived by the "yoke of patronage" of their sacred, scriptural right of calling their own ministers. The new act, argued the pamphlet, extended patronage beyond the legal requirements in not allowing the moderation of a congregational call until six months had passed or the patron had consented. Moreover, the new act granted all heritors, resident or not, Presbyterian or not, moral or wicked, an equal vote with the elders to the exclusion of the people. The authors spoke most strongly against settlements made upon presentations when the congregation objected and against the acceptance of such presentations as violent intrusions: "coming in without the Call or consent of the Christian People, and as having their own carnal Interests more in View than the Glory of God, and Salvation of Souls." Furthermore, the authors of the *Testimony* declared that they would, conversely, have to support honestly called ministers; and if this resulted in schism, the assembly was at fault for creating an unconscionable situation. What was most interesting, however, was the failure of the authors to be satisfied with their statements against patronage. They continued to chronicle all the church's recent sins, including the abjuration oaths, the inadequate trial of Simson, the condemnation of the *Marrow*, and the great sin, neglecting the Solemn Covenant.[14]

Through the following year the Presbytery and the Synod of Perth and Stirling tried to make peace, to no avail.[15] In December 1733 the four ministers declared themselves a formal judicature, the Associate Presbytery. They recorded eight justifications for their action. As Presbyterians they were obliged to form such an association; in order to maintain discipline a formal judicature was necessary; they hoped to display "the Testimony put by the Lord in their hands"; having been cast out, it was their duty and profit. Perhaps the most important reason from their viewpoint was the last: "The present case of the Established Church, their hastening to ruin the Lord's Work. . . ."[16]

The assemblies of 1734 and 1735 attempted reconciliation, releasing the sentences and repealing the resolution of 1730, but the Associate Presbytery judged these approaches too little, too late. The vast sins of the assembly still remained unrecognized; thus they remained steadfast, or obstinate, depending upon the point of view. The General Assembly described them as arrogant schismatics.

> With the Air of a paramount Power and Authority [they] condemn this Church and the judicatures thereof . . . and [they] cast many groundless and calumnious Reflections upon her and them.

They had erected themselves into a presbytery over the whole church, a power that no association of ministers had ever claimed against the assembly. After attempting reconciliation with the Seceders, albeit on the church's terms, for several years, the Church of Scotland brought in a libel against them, and as they refused to appear (except once in 1739 to deny the assembly's jurisdiction) they were deposed.[17]

The Associate Presbytery had no reason to rejoin the Church of Scotland, since it was eminently successful in its own right. The General Assembly complained of the vast number of adherents.

> They appoint and keep Fasts in different Corners of the Country, to which there is a Resort of several Thousands of Persons of both Sexes, and too many of them, as there is good Ground to think, come there with other Views than to promote Religion. . . .[18]

Whether the goals of the people were spiritual or otherwise, they did seem to flock to the relatively small organization of four ministers (seven by 1739). In December 1735 the Associate Presbytery received its first request for supply from a congregation that had left the jurisdiction of the Presbytery of Auchterarder because the assembly insisted that they defer to the patron in the choice of a minister. In January 1737 the Associate Presbytery received a petition from the "General Meeting of the Societies in the South and West of Scotland" seeking supplies for several "praying societies" or seceding congregations. In May 1739 they received petitions from twenty-four congregations who had seceded, sixteen of them without the assistance or persuasion of any pastor; that June thirty congregations sent requests; in December, thirty-three; and the following July thirty-seven separate congregations made some request of the Presbytery.[19] Many people had at last found an ear sympathetic to their confusion and dissatisfaction.

Were the congregations urged to break with the General Assembly? Did the Seceders play upon their emotions of fear and zeal, or were that many people independently disgruntled with the status quo? The congregation of Muckhart needed no encouragement to feel ill-treated. The patron of Muckhart had, in May 1732, presented Archibald Rennie to those living there, an appointment that was supported by only four heritors (two of whom were not resident) and opposed by the session's three elders, seventeen heritors, and the entire congregation. The Presbytery of Auchterarder rebuked Rennie for accepting the presentation and asked him to withdraw; he refused and the process of installation was delayed for a year. The following March the presbytery allowed the moderation of a call and allowed Rennie a place on the leet, or ballot. The congregation, elders, and majority of heritors called John Haly, and the patron's commissioner protested any action but the calling of Rennie. The commission of the General Assembly in August of 1733 ordered the presbytery to moderate in a call to Rennie alone; Auchterarder refused, and the commission itself decided to moderate the call.[20] The commission heard Rennie's trials in March 1734 and settled him in the congregation, but the presbytery refused to accept him as a member, juggling excuses and procedures for several months to avoid accepting the intruder. The August 1734 minutes recorded that Rennie occasionally had only three or four hearers; the congregants petitioned to be permitted to attend other parishes. For four more years the presbytery fought the synod and General Assembly over the enrollment of Rennie as a member. As of 1745 he had not been accepted by the presbytery.[21] The opposition remained so strong that he preached only once in the church building, his first Sunday. He never had an elder, never distributed communion, and led small services in the dining room of the manse.[22] This trouble was certainly not aroused by rhetoric but represented a real grievance held by a congregation against the church government. The opportunity offered by the Associate Presbytery for the congregation to reassert its traditional role in the decision-making process was too tempting for an angry, affronted group.

The records of the Session of Perth, the congregation served by Wilson, one of the original Seceders, tell a more complicated, less extreme story. During the summer of 1733 the congregation had joined the Presbytery of Perth and Stirling in the request that Wilson and the others not be suspended, for their congregation would be unable to function if he could not practice his ministry. A commission was slated to deal with the ministers in December, so a petition for delay was presented in December; it was refused, and the ministers were suspended. For three years there was no more notice taken of the assembly's actions regarding the Seceders except that Wilson, supposedly suspended, continued to moderate the session and hold communions. Suddenly in the summer of 1737 Wilson objected to the new minister installed at Perth, David Black, a member of the established church. He could not sit in session with him or even accept that such a meeting was a session. He would (and did) leave them and constitute a *real* session with those members who felt as he did, and he branded his opponents as guilty of seceding. With Wilson went several elders, including the clerk. A statement of the reasons given by Wilson for his actions no longer exists, but the response prepared by the remaining elders attests to the character of the event.

The session declared that Wilson's departure made sense since he and other ministers had already seceded from the church several years ago; however, Wilson's refusal to commune with David Black because he was a member of the national church was sheer pretense, since Wilson was content to sit with Thomas Black, also a minister of the Church of Scotland. Apparently Wilson accused David Black of being a patronage appointment, called by the town council only, but the session asserted that this was not true. Black was legally called. The elders, by a majority of one, may have voted against him, but the congregation had voted for him by a large majority. The Seceders themselves claimed that elders had no more right than any other group in the congregation to determine the choice of the minister. Finally, the elders accused Wilson of kindling division within the congregation. For more than a year after Wilson's minisecession the Session of Perth tried to make peace among the factions, but the new organization was apparently content in its separation. In December 1738 the session held elections to replace the seceding elders, and the division was buried under the more pressing immediate business of the church.[23]

More than patronage was at stake. The Seceders believed they were fighting for the life of Scottish Protestantism. In 1736 Archibald Campbell, author of *An inquiry into the origin of moral virtue, The Apostles no Enthusiasts*, and *de Vanitate Suminis Naturae*, was called before the General Assembly to defend some injudicious expression in his works. Although the assembly dismissed him without censure, the Seceders found his works contrary to the Word of God, full of "gross and dangerous errors" and "a Manifest Tendency to subvert all Religion both Natural and Revealed."[24] The Seceders saw the assembly's action as following an inclination on the part of the established church to selectively prosecute challenges to orthodoxy. Suggestions of antinomianism were quickly and thoroughly rebuked, while outright assertions of Arian or Arminian tenets were lightly admonished with a mild suspension. The Seceders could not forget that the assembly had never deposed Simson but merely let the matter fade from the public view while he was still under suspension.

Nonetheless, there was more to the Seceders than strident opposition to supposed abuses. In its *Act, Declaration, and Testimony*, the Associate Presbytery outlined systematically its theology and views upon church government. The declaration reads like a sixteenth- or seventeenth-century Calvinist creed. If these ministers truly believed their declaration, it is no wonder that they felt threatened by the movement of theology during their lifetimes. A benevolent deity and striving humanity were not compatible with their traditional Calvinism.

The Seceders believed that nature demonstrated the existence of God and that scripture was the only revealed rule of life. Also included was a declaration of the integrity of the Trinity and deity of Christ. From these basic truisms the declaration continued to the eternal nature of God's unchangeable decrees and an affirmation that these decrees were made by grace without foresight of faith or works. In other words, salvation was completely the work of God and not the result of any human effort. Original sin was imputed to all mankind, bringing a total depravity that rendered ridiculous the pretensions of the Arminians. The light of nature was simply not sufficient to provide the knowledge of God and God's will necessary to salvation. Only the incarnation of God as man gave surety for the elected sinners. Christ yielded perfect obedience to the law and bore the wrath of God as the true sacrifice. Christ's righteousness was imputed to the elect, and the grace of faith became the instrument, though not the grounds, of salvation. All sin deserved the curse of God; the only measure of virtue was the law of God, and all believers were forever under the law. However, God was all-sufficient. He voluntarily promised rewards to the elect, and those who believed could not again fall totally from grace. There was no room for human efforts, virtues, or wisdom. All were dependent upon the will of God.[25]

Theirs was a harsh creed, inflexible and destitute of hope for the unconverted. Yet in the place of hope was an assurance lodged within a belief in the omnipotence of God. That humanity could do nothing also implied that God did do everything. While stopping short of a proclamation that an individual was not responsible for sins committed (a proclamation that the progressives frequently claimed the evangelicals believed), they did clearly believe that the actions of God had provided the security of humanity.

From this vantage point the Seceders' position on civil interference in church affairs made perfect sense. Christ was sovereign over the church, and the Bible provided the only rule that Christians were obliged to follow in organizing their societies. Thus the Seceders rejected four "Erastian" principles: that the civil magistrate was supreme; that the officers of the church in their spiritual capacity were subordinate to the magistrates; that the government of the church depended upon the civil magistrates; and that the civil government might pass acts to govern the church. These were all rejected in favor of the position that the presbytery was a divine institution ordained in the New Testament; and only by virtue of one's adherence to the Christian faith did any individual earn a voice in the governing process of the church. The Seceders even condemned the General Assembly for allowing the civil government to name the day of the national fast. "There is a Sinful combination between Church and State, to make an Invasion upon the Headship and Sovereignty of our Lord Jesus Christ over his Church as his free and Independent Kingdom."[26]

Was this belief system truly appealing, or did the congregations merely rally round the Associate Presbytery because it gave them an opportunity to virtuously oppose and successfully defeat the authorities? Whatever else might be said, the Seceders were zealous in any cause they supported, and zeal seemed to attract many believers. When the Associate Presbytery declared a fast, they were unrelenting in their goals and accusations:

> unbelief, neglect of the great Salvation among Gospel hearers, open Atheism, Unfidelity, Despising the Lord Jesus Christ, and contempt of his Glorious Gospel: —Lukewarmness, Security and Neutrality in the cause of Christ among Professors: — Impenitency, the hardness of heart under the marks of the Lord's displeasure: — Open profanity, disregard to the Lord's day: Uncleanness, and innumerable abominations of all sorts. . . . And considering the great and manifold indignities done against the Lord Jesus Christ, his Truths and Interest, in these Lands, contrary to our solemn engagement to the Lord by our Covenant both National and Solemn leagues not only in former times but of late; particularly that the most high God our Saviour hath been blasphemed; and his proper true and Supreme Deity impugn'd and denied.[27]

Concerned primarily with purity of doctrine and godliness of conduct, the presbytery managed to touch upon most aspects of sacred and secular existence. Its call for a fast read much like calls for fasts everywhere: catalogues of vices and sins, many of which (e.g., lukewarmness, hardness of heart) were commonly cited by all kinds of churches during this century. It almost seems as if anyone could have produced that indictment of their society. One item, however, immediately identified the source of the accusation. Toward the end lay the reference to the covenant, a sacred institution that the Seceders, following the evangelical example, refused to give up.

The Seceders were ritualists, traditionalists. They stuck by a version of Calvinism that was rigid and clear. They fought against the civil government and, in the end, the church government, defending the spiritual independence of the church. They called for a strict adherence to the austere morality of John Knox and the sixteenth century. They believed that they put the Bible foremost in their understanding of God, but in actuality they put the traditions of Scotland foremost. They fought progress because it was progress, because it marked the end of Scottish tradition and the transformation of the church into an institution that had no traditions. Unwilling for whatever reason to explore the speculation kindled by the Enlightenment, the extremist evangelicals defended their religiosity stridently, invoking the only precedents they knew and understood, the traditions of their church. Every statement published by the Secession Church, every call for a fast, every accusation against the General Assembly began with a historical account of religion in Scotland. From John Knox to their own movement, the Seceders identified a single tradition of pure, reformed religion to which they had pledged themselves.

The *Act, Declaration, and Testimony*, published by the Associate Presbytery in 1737 as its definitive call to arms, was a wonderful construction of this tradition. The introduction, a small but effective piece of vituperation, emphasized that the General Assembly had cast out the Seceders and that it had become the Seceders' duty to form a presbytery for the relief of congregations who had suffered intrusions. By their example they were giving doctrinal and judicial testimony and yet were condemned as guilty. The presbytery had watched while the General Assembly accepted more presentations and passed over Archibald Campbell without censure. They had waited three years for the General Assembly to reform and, when it was clear that it would not, the Seceders had decided to declare themselves an official presbytery and make a public declaration of faith.

The public declaration began in 1560 with John Knox and the Book of Discipline, traveled quickly to the outrageous Acts of Perth, and then enjoyed a brief respite in the reformation of 1638, "the purging the House of God." It proclaimed the revivals in Shotts and western Scotland as the continuation, despite prelacy, of the Protestant spirit. The 1640s were years of great gospel success. England and Ireland had joined the Solemn League and Covenant, binding both nations to personal and national reformation, and the Westminster Assembly had devoted four years to the development of a masterful program of orthodoxy, discipline, and church government. Unfortunately, great defections were to follow. Opponents of the Solemn League, upon a formulaic repentance, were returned to office. With the restoration of Charles II all the acts of the Long Parliament were nullified, and prelacy was reestablished. Supremacy of king over church was again the law, gospel ministers were deposed for nonconformity, and the Scots endured persecution and immorality. Pastors "deserted" their congregations to prelatists. House and field conventicles were persecuted, and, during the time of restrictive tolerance, some conventicle preachers connived with the authorities in abiding by the conditions of the official indulgence.

As they continued to list the numerous sins of the English government during this period, the purity of the tradition became clear. Their heroes were those leaders who took an extreme Presbyterian position and refused to waiver. They held out for absolute, essential, unblemished Presbyterianism. Up until 1688 there had always been a group identifiable as the radical Presbyterians. These were the ministers who had always insisted upon independence of the church government in its presbyterial form; rigid, austere morality; and the need to maintain the purity of the ministry. Frequently they fought the acceptance of penitent, recanting pastors and members with harsh accusations and grudges for past offenses. But these were the ministers who traditionally had appealed to popular support. Their tradition included not only the refusal to conform to the Acts of Perth, but the 1630 revivals. They had not only protested prelacy but kept the faith alive in conventicles during the Restoration. They had actively sought popular support and reaped the benefits of their efforts. More than a self-fulfilling prophecy, the attraction of numerous individuals to the radical Presbyterian cause was a concrete, observable phenomenon. Initially the people were brought to the church through a system of beliefs and rituals that served popular spiritual needs. In the end these rituals served religious needs because they were popular traditions. As one sympathetic critic noted, "If they erred in anything, I apprehend it was in consulting human writings and human commentaries too much, and giving too little scope to their own independent researches, reflections, and discussion."[28] They found their strength in their connections with that known quantity, the past.

Whether self-conscious or not, the Seceders were strict ritualists. They placed emphasis upon forms and ceremonies, prescribing proper procedures for every action. Being staunch Protestants, the Seceders condemned most recognized rituals as unspontaneous and superstitious. They refused even to read prayers from books during a service; rather the pastor prayed spontaneously for an hour or more. They rejected all hymns and paraphrases of the psalms, satisfied to read the psalms plainly. Yet for all their efforts, they, like so many Calvinists, ended by making a ritual out of plainness and austerity and by allowing such rituals to dig deeply into all aspects of religious life.

Excellent examples of this type of ritualizing could be found throughout the records of the Associate Presbytery. In one case, unable to leave Simson alone, the four seceding ministers lamented the assembly's light sentence when it was clearly proven

> that he had Openly and Wickedly blasphemed the Son of God by impugning his Supreme Deity, Independancy, necessary Existance, and Numerical Oneness with the Father, and by asserting that the title of the only true God might be taken from the Person Property of the Father and so [it was] not belonging to the Son. . . .

This accusation formed part of a confession by Alexander Moncrieff that he did not respond properly to the crisis. The law of God directed "that Persons who blaspheme the Son of God, in the manner he had done without convincing evidence of their repentance should be delivered over to Satan," but Moncrieff did not even protest the assembly's decision of a year's suspension, and now he begged the forgiveness of the presbytery. Erskine confessed that although he had adhered to the protest he had sinned in not insisting that such protest be recorded. Not to be outdone, James Fisher confessed that although he was not a member in 1729, not only did he not protest but he did not join in the testimony offered afterward. The penitent ministers then had the pleasure of censuring and forgiving each other. The case of Wilson was truly ridiculous. He confessed that although he was not a member of that assembly and was not present, yet, when the assembly's decision was reported in synod and presbytery he did not protest it. Even the rigid Seceders could not admonish Wilson since he had joined the testimony as soon as it was prepared. Thus wallowing in these admissions of minute sins of defection, the four ministers would guarantee the purity of their church as compared to the laxity of the establishment.[29]

For men opposed to the merest hint of any prayer-book form, the Seceders indulged in the most stately of rituals, the renewal of the covenant. Ministers would advertise beforehand with weeks or months of preparatory sermons and instruction. Often the ritual would take place on the fast day preceding the communion, linking temporally the two Scottish traditions. After a special sermon had been delivered, the names of all who had decided to subscribe were read, and those individuals then stood physically apart. The covenant was read, followed by an "Acknowledgement of Sins and the Engagement to Duties." There followed confessory prayers, a psalm, the administration of a sacred oath, and a special exhortation to the jurants. Finally, the congregation would break out in joyous psalm singing while the covenant was signed.

> The service was very impressive, many of the audience were occasionally in tears, and the result was usually a deeper-toned religious life, a revival of the

work of grace in the congregation, and the strengthening of that high-toned
religious principle for which Original Seceders have oftentimes been distinguished.[30]

In all fairness to the Seceders, they had committed themselves to more
than a document designed a century before to resolve a specific political
problem of alliance. Somehow understanding that spiritual life was the
appropriate goal to be sought, the Seceders had designed all their rituals
around the Holy Spirit. They had ritualized plainness because they valued
universal understanding and spontaneity. Metaphor and eloquence were
possible only by thorough preparation, and this might preclude the work of
the Spirit within. Like their traditional ancestors Blair and Livingstone,
they rejoiced in extemporaneous homilies that they believed were directed
by God.

> Their delivery was sometimes uncouth, but always earnest; their language
> necessarily, and even of a purpose . . . was unpolished. Their discourses,
> though not very compact, were replete with gospel truths and scriptural
> references, which last were generally quoted chapter and verse.[31]

When they first began to consider candidates for the ministry, the
Associate Presbytery examined the probationers in "their experience of the
power of Religion and Godliness upon their own Souls," in addition to
their erudition and preaching abilities.[32] Although their major concern
throughout the eighteenth century would be the purity of the Christian
communion, the Seceders had begun to translate some of those concepts
onto the personal level. The purity of the ministry was definitely a community imperative, but ministers were individuals, and some of the concerns expressed and resolved for the ministerial candidates eventually
became the presbytery's concerns for individual members.

> Cherishing great seriousness of soul in matters of religion, these exercises
> were intense and scrutinizing, and they were apt . . . to fix the evidence of
> personal religion higher than Scripture examples and intimations would
> warrant and the scheme of grace would justify.

In their efforts to bring individuals to their point of view, the Seceders were
inclined toward lengthy and vehement argument.

> They were apt also to appeal to their feelings and frames more than was
> meet, and to resort to these as the criteria of a personal interest in Christ,
> rather than directly to the Word of God and the testimony of Jesus.[33]

They actively sought converts, assisting each toward his or her own conversion experience, appealing to loyalty to the past, plain persuasion, pious

emotion, whatever would serve to bring the individual into the true Christian fold. By these dual efforts to call individual believers and to awaken the community as a whole, the Associate Presbytery worked toward the reformation of the Scottish nation.

News of the Secession traveled quickly. It has already been told how different communities in southern and western Scotland organized themselves into praying societies and joined themselves to the movement. And in 1736 the Associate Presbytery received its first correspondence from the congregation of Lisburn in northern Ireland complaining of the imposition of a minister and requesting supply of preaching.[34] That Ireland would have heard of the movement so quickly was to be expected, since Scotland had remained the primary locus of formal education for Irish Presbyterian ministers, effecting the continuation of lines of personal communication. That the Scots-Irish were so unhappy as to continue to push toward schism upon schism was just as predictable, although the causes were a bit more complex.

The subscription controversy had left its scar upon the Irish Presbyterian church. Some ministers had been cast out of the communion, against the wishes of most of the Irish clergy. Friends found themselves members of different communions; the peace and unity that had been so anxiously sought and carefully preserved was, in a few years, formally destroyed. Nonetheless, many clerical friendships remained strong, and the regular attendance of some nonsubscribing ministers at the synod to discuss the distribution of the regium donum served to maintain those ties of friendship. The nonsubscribers and the Southern Association, along with many clerics sympathetic to their views, had hoped that this continued contact would defuse the hostility and enable the two groups to reunite in a single communion. Unfortunately, the product of this clerical conviviality was not defusion but confusion.

Although supported by several strong, conservative ministers, the laity had actually been the force behind the schism. They wanted to maintain the purity of the church. When they observed their traditions threatened by the erudite speculation of the Ulster intellectual community, the laity actively fought that influence. Having before them the example of Salters' Hall and operating under the misapprehension that religiosity equaled theology, the conservative party had turned to subscription as the guarantor of orthodoxy. The laity soon discovered that subscription guaranteed nothing except the ability of a minister to perceive his own beliefs as agreeable to the church's standard, the Westminster Confession, and this was not enough. So what was enough? The problem was that no one knew. The years of congregational hostilities and defections that followed the schism of 1726 were ample testimony to the confusion that governed the progress of Presbyterianism in Ireland for most of the eighteenth century. The people were seeking some remnant of the true reformed church. They were not sure how to recognize the true church; indeed, different congregations would latch onto different aspects. However, the people were certain that so far a pure Christian communion could not be found within the establishment.

The mainstream persuasion was only one example of popular discontent that resulted in the formation of more traditional, evangelical congregations in direct competition with the contemporary organization led by nonsubscribers. The very suggestion of nonsubscribing principles usually sent a majority of the people to other congregations or, failing that, to petition for the erection of a new congregation. Some congregations did remain loyal to their pastors though disagreeing with their principles. For example, the people of Ahoghill stayed with Thomas Shaw, installed in the church in 1710, and joined the nonsubscribing Antrim Presbytery with him in 1726. However, after Shaw's death in 1731, the congregation rejoined the General Synod.[35]

The laity were not prepared for change as drastic as nonsubscription. The congregation at Antrim had provided a pulpit for John Abernethy, one of the leaders of the progressive party. The people had been devoted to their pastor, fighting the very Synod of Ulster to keep him. In 1717 a delegation from Usher's Key in Dublin presented a call to Abernethy. As members of the Dublin Presbytery of the Southern Association, the commissioners promised to submit themselves to the Synod of Ulster if Abernethy were transported there. The synod voted to keep him at Antrim. The following year the delegation from Usher's Key returned, along with one from a Belfast congregation, both soliciting Abernethy's services. The synod changed direction, voted to transport Abernethy to Dublin, and therefore severed his ties with the Ulster Synod.

The people of Antrim protested that they still had a claim. The next year the moderator asked whether Abernethy had obeyed the synod's decision. He replied that he had acquiesced regarding the ending of his relation with Ulster, but as he was no longer a synod member, the synod could no longer call him to account, and he chose to stay with his congregation, who still wanted him. The entire issue was rediscussed, with the congregation from Dublin now claiming that they had a right to Abernethy by the synodical decision, and that the synod could not revoke that decision without their consent. Abernethy was requested to give Dublin a trial, and the presbyteries were warned against concurring in any call (specifically Antrim's call) to Abernethy without his release from Usher's Key. The meeting of 1720 revealed that Abernethy had steadfastly refused to supply Dublin, and finally Usher's Key had released him from all obligations. The synod grudgingly permitted him to remain at Antrim.[36] These people had evidently stood by Abernethy, refusing to let him go despite the order by the General Synod. Yet in 1727, after Abernethy, as a member of the nonsubscribing party, had been ousted from the synod, the majority of his congregants petitioned the synod for a new congregation under the discipline of the General Synod.[37]

If the specters of unorthodoxy and false religion had merely plagued those congregations with nonsubscribing ministers, Irish Presbyterianism might have survived almost unscathed. But the infection had spread beyond a caution of those self-confessed defectors to suspicion of any minister who had sympathized with those principles or even with the nonsubscribers as individuals. Presbyteries began to take on Old Light (conservative) and New Light (progressive) characters. Killyleagh was a fairly flexible presbytery; one of its members had kept communion with Alexander Colville in 1726 and 1727.[38] The congregation of Drumbo asked to be removed to a different, more orthodox presbytery.[39] The Presbytery of Bangor, formed as an association for the subscribing neighbors of the Antrim Presbytery, was the progressive counterpart to the orthodox Presbytery of Templepatrick. In 1743 the Presbytery of Dromore was created to satisfy several unhappy congregations under the direction of the liberal Presbytery of Armagh.[40] Some of the presbyteries began to play partisan roles, so that congregations were occasionally able to play one presbytery against another to achieve their goals.

What were their goals? As the century progressed it was increasingly apparent that the congregations did not know. When William Cunningham, pastor of Limavady, died in 1740, the congregation immediately divided over selection of a new minister. The wealthier members favored Joseph Osborn, a repentant nonsubscriber. Since one of Osborn's supporters owned the meetinghouse, his party cast from the building the majority under the leadership of the presbyterially ordained Henry Erskine. Erskine and his followers were obliged to meet in the open fields.[41] The Synod of Ulster in 1747 removed William Laird, against his wishes, from his pastorate at Ray to the more prestigious congregation in Belfast. The people of Ray turned to the incoming Secession Church, seized the meetinghouse, and refused access to the supplies appointed by the Presbytery of Letterkenny.[42] When the pulpit of Ballynahinch became vacant in 1733, the congregation split their vote between two candidates, with the supervisory Presbytery of Bangor refusing to ordain either one without a more general concurrence. The synod finally annexed the majority party and their candidate McClain to Killyleagh, leaving the minority, who were "not edified by his sermons" to join other congregations in Bangor.[43]

One of the saddest situations during this turmoil was that of the congregation of Lisburn, whose pastor died in 1731. Lisburn had always been a problem; pastor Alexander McCracken had been one of the loudest, nastiest opponents of the nonsubscribers during the 1720s. The minutes of the Lisburn Session displayed the full panoply of hostility that ran through this congregation. Accusations of thievery, lying, quarrelsomeness, and fraud were frequently thrown about. McCracken and some of his elders were involved in a gambit of attacking and salvaging reputations. After 1720, however, the congregation had united behind McCracken's leadership. The threat of nonsubscription had galvanized the congregation into acceptance of a leadership role in the struggle for evangelical integrity.[44] It should not have surprised anyone that after McCracken's death some members of Lisburn would remain allied to the staunch conservative cause. The majority called Gilbert Kennedy, certainly a strong, published advocate of subscription; but a significant minority questioned Kennedy's orthodoxy. Then began a series of calls by each side to its own candidate. The two candidates would be set aside by the presbytery and the congregation instructed to try again. After five years and four such attempts, the Presbytery of Templepatrick finally ordained William Patton of Monaghan. The minority immediately protested, and, though granted permission to attend other congregations, they complained to the Associate Presbytery that a minister had been forced upon them by their presbytery and petitioned for supply.[45]

The problem of patronage was totally irrelevant to the Irish Presbyterian community, since they were a dissenting organization rather than an established church. The Irish congregations had since 1642 enjoyed the privilege of calling their own pastors by popular election, while the Scots had always delegated that responsibility to the elders and heritors. Supposedly the Ulster Irish had settled all suspicions of unorthodoxy in 1726 when the synod had ousted the nonsubscribers, but this clearly was not so. In the midst of their confusion many lay persons saw in the Secession movement that remnant of the true church that they had been seeking. Certainly they would be attracted to a ministry that allowed them, even approved their intention, to do as they pleased, whether the church authority supported or opposed them.

However, the Seceders did not advocate a laissez-faire policy regarding their churches. The Associate Presbytery made strenuous demands upon a congregation before taking the body under its care. In the case of an early petition from the congregation at Markethill, the Associate Presbytery advised that the people devote more effort to their understanding of the *Act, Declaration, and Testimony* and then consider petitioning the presbytery. The congregation responded with a carefully worded request constructed within the forms of the published document, and they were accepted as a mission congregation.[46]

Despite this pedantic devotion to their forms and structures and the absurdity of their cause—antipatronage—in the Irish situation, the Seceders enjoyed enormous success. After the first congregation was established under Isaac Patton in 1746, the number of Secession congregations multiplied at a tremendous pace. In 1748 the Associate Presbytery split again over the propriety of taking a specific magistrate's oath. While this division had absolutely no meaning in Ireland since that oath was not enforced there, the competition between the Burgher and Anti-Burgher synods for congregations and ministers increased the efforts of both groups in Ireland. By 1760 over twenty Secession congregations had been established in the north; by 1780 the number had more than doubled, and both the Burgher and Anti-Burgher organizations had established Irish presbyteries.[47] Partly this spread could have been due to the unwillingness of the Ulster Synod to create new congregations because of the scarce financial support available; partly it was due to the Seceders' encouragement of congregational discontent and division. Mostly the laity was attracted by the quality of the Seceders' religiosity.

In 1743 the pastor of Markethill, George Ferguson, was accused of denying the doctrine of original sin. Many of his congregation refused to attend his ministry further. The orthodox Presbytery of Dromore began to supply the malcontents so that their righteousness might be preserved. This occurred without the concurrence of the Armagh Presbytery, which immediately protested the intrusion. The General Synod censured Dromore and called for an immediate investigation of Ferguson. Since no witnesses testified to the court of inquiry, the verdict could only be "not proven," and the discontented remained discontented. They were finally granted permission to transfer to the care of Dromore, but were instructed to join existing congregations rather than form a new one. In despair they appealed to the Secession Church and, after learning the tenets of that particular system, were granted supply and in 1749 created a congregation with pastor.[48]

When the Seceders first preached in Ireland during the early 1740s, they expected to be welcomed by the establishment. The Associate Presbytery had observed Ireland's subscription controversy and believed the result a victory for orthodoxy. Thus, they expected to join forces with the Synod of Ulster against the nonsubscribing heretics. At first welcomed by the more evangelical members of the synod, they were soon feared for their schismatic abilities. Actively instilling zeal for truth and confidence in their capability to discern that truth, the Seceders traversed the countryside calling for all true Protestants to join them and cast off hypocrisy. In the end the synod's response was as much self-defense as anything.

A Serious Warning to the People of our Communion within the bound of the Synod was the response published by the General Synod in 1746. It began with a list of theological errors that were creeping into progressive theology, including the denial of the doctrine of original sin and the real satisfaction provided by Christ; the emphasis upon obedience to the moral law to qualify believers for communion with God; and general skepticism concerning gospel ordinances and disbelief of the scriptures. They also advised that erroneous books be kept from unlearned individuals. The majority of the statement, however, was formulated against the Seceders, who were traveling the countryside railing against the heretics who abounded within the ministerial ranks. The synod warned its members against the Seceders as fanatics and schismatics; hurling accusations was a most unscriptural method of preserving the truth. The synod, as they reassured the laity, carefully investigated and tried any member suspected of error. If anyone, clerical or lay, suspected a minister of unorthodoxy, a libel should be presented, and the man would be tried. All judicatures were encouraged to preserve the truth, and the presbyterial courts were the proper forum for such accusations. The synod found that the Seceders' practice of blindly railing against regularly ordained ministers in their path was divisive and disastrous for the progress of the gospel cause in Ireland.[49]

Despite synodical hopes and warnings, the Secession flourished in Ireland and in Scotland. In Ireland it had little competition. Had the Ulster Synod acceded to more congregational requests, the tale might have ended differently. For example, when the Ardstraw pulpit was vacated in 1729, the congregation split into two factions that, for three years, considered several candidates without agreement. Finally, after four probationers had been heard several times each, a poll was taken, and, because the opposition failed to vote, Andrew Welch was selected pastor and ordained by the moderate Presbytery of Strabane. For six years the minority opposition refused to support Welch's ministry. In 1739 it was discovered that a member of the evangelical Presbytery of Letterkenny was holding meetings for those dissenting members. Although nothing could be proven against Welch, even by his strongest opponents, and most were content with his ministry, a significant portion of the congregation found him not quite right. In this case, perhaps because of the large original membership of Ardstraw (five hundred families), the synod allowed a new congregation to be formed under the supervision of Letterkenny, and the trouble ended.[50]

Nevertheless, the Synod generally opposed too strongly what they viewed as the proliferation of congregations, and the ministers continued to reject the criticisms of the laity. They would insist that the church was the proper business of the clergy, that lay persons were not competent to have a reasonable opinion, much less exercise influence. Thus many Irish congregations turned to the Secession Church, not so much because the Associate Presbytery heeded their advice, but because their theology and practices matched some image of the true, reformed tradition and therefore better fulfilled their spiritual needs.

In Scotland, however, the Seceders did not proceed without a contest. George Whitefield, fresh from his success in British North America, was invited to Scotland in 1741 to preach the gospel. The initial invitation came from the Associate Presbytery; they saw in Whitefield a kindred spirit concerned with the spread of true religion. Whitefield, though espiscopally ordained, engaged in a religious dialogue outside the purview of the defective Church of England. However, when the two voices of popular religion met and talked, they found themselves in violent opposition. The Associate Presbytery had invited Whitefield to address its own congregations, and asked Whitefield to restrict his activities to its communion. When Whitefield asked why, the presbytery responded that its congregations represented the only true church in Scotland. Whitefield, who never had any intention of restricting his audience anywhere, replied that his business was generally with sinners. If the members of the Secession Church were already saved, then his time ought to be spent preaching to others. The Associate Presbytery performed an abrupt switch.

> The Presbytery taking under their consideration the Reception of Mr. George Whitefield a Priest of the Church of England has met with in Scotland contrary to our Covenants whereby Episcopacy and all that order are abjured, with the consequences of said reception, particularly these appearing at Cambuslang and other places; and considering the awful aspect of that Providence and the many evidences of a Spirit of Delusion which God is permitting to go through this Land . . . [51]

the Associate Presbytery decided to hold a fast.

The prayers and fasts of the Seceders notwithstanding, Scotland was rocked by popular revival.[52] Once again the western and border areas experienced the highest incidence of popular outburst. Campsie, Kilsyth, Calder, Cumbernauld, and Glasgow were among the towns listed as reaping the benefits of revival. James Robe's *Narrative* of the events at Kilsyth, though formulaic, provides some useful discussion. He had been pastor of Kilsyth since 1713, originally finding the people well educated and committed to religion. In the winter of 1732–33 a pleuritic fever swept the village, taking some sixty inhabitants in three weeks, including many pious and judicious Christians. From that point on the state of religion declined, unawakened by storms of hail and thunder. Then came the Secession, causing divisions among the people, many of whom sympathized with that cause.

> They were so bewitched as to incline to the separating Side, and were so taken up with disputable Things, that little Concern about these of the greatest Importance could be observed among them. All Societies for Prayer were then given up.

More storms and a drought went unnoticed. In the summer of 1740 Robe began a long sequence of sermons on regeneration that, with some interruptions, ended in the spring of 1742. He had heard of awakenings at Cambuslang and prayed for his own congregation. At last in May his village was stirred. Within a four-week period, Robe reckoned that perhaps over two hundred parishioners and another hundred visitors had been affected. Robe was quick to add that this reform had been reflected in behavior as well as in emotional responses at meetings.[53]

The reports of the work at Cambuslang followed much the same form. The preaching was focused upon regeneration. The initially converted corps numbered fifty, and the size increased daily. Ministers delivered five or six sermons weekly and were involved in constant private conferences. Every favorable pamphlet noted a true change of heart among the converted: visible reformation in the lives of notorious sinners in a lessening of cursing, swearing, and drinking; remorse for acts of injustice; public restitutions; forgiving of injuries. One reported an increase in family worship and a new love for the scriptures and ordinances. There was "extraordinarily *fervent* Prayer in large Meetings," but pains were taken to discover the hypocrites.[54]

No matter how the ministers tried to emphasize the content of sermons and the permanent moral reformation, there was no escaping the physical enthusiasms of those revived. Alexander Webster turned these potential embarrassments into virtues when he described the tears, tremblings, cries, and swoons. "Nor does this happen only when Men of *warm Address* alarm them with the *Terrors of the Law*, but when the most deliberate Preacher talks of redeeming Love."[55] Robe perceived the physical effects as a predictable result of spiritual awakening. "I found the Distress of their

Minds to be so great, as they could not but naturally have such Effects upon the Body."[56]

The opposition ran amuck with the physical outbursts of the congregations. One attack specifically linked the revival at Cambuslang with the activities of a shoemaker who, after hearing Whitefield in Glasgow, began to hold private conferences,

> stirring up Convictions in a great many, and by the lively Description of his own Feelings (which he was at all Times ready to give . . .) wrought up Several to an earnest desire of the same pleasant Sensations.

After three weeks of such conferences, some people began to cry out

> saying *they now saw Hell open for them, and heard the Shrieks of the Damned* and express'd their Agony not only in Words, but by clapping their Hands, beating their Breasts, terrible Shakings, frequent Faintings, and Convulsions, the Minister often calling out to them, *Not to stifle or smother their Convictions, but encourage them.*

Conversion was portrayed as immediate and complete—"they are generally rais'd of a sudden, from the deepest Agony and Grief to the highest Joy and Assurance"—and the converted as arrogant, presumptuous fanatics who went about preaching from the rooftops and giving advice to the still unconverted. As the service began, the distressed souls would sit at the front of the tent with dishes of water for the fainters; they cried through the entire service: psalms, sermon, and prayers. The original shoemaker was quoted as saying, "Lord, DUNG us with Jesus Christ," and "A Woman who had been formerly a Seceder, declared openly before Mr. *Fisher* and his Elders that she had seen Christ with her bodily eyes."[57]

The defense against such attacks was quick and diverse. James Robe, in four different pamphlets addressed to James Fisher, returned to his assertions of permanent reform and issued a mild challenge to the Secession Church's arrogance in restricting the potential sphere of divine activity. Citing Ralph Erskine's *The True Christ no New Christ*, "New and strange Sort of Convictions are to be suspected if, instead of Convictions, we hear of Convulsions, bitter Outcryings, Frights, Faintings and Foamings. How delusive is the Work," Robe responded that the phenomena themselves were not the works of the Holy Spirit, but the work of the Spirit might produce such results. Did the Seceders dare to limit the actions of the Spirit? A single conversion was the work of the Holy Spirit; certainly the Spirit could convert more than one at a time.

Familiar with the activities of the Seceders, Robe sampled Ralph Erskine's sermons and found compositions more replete with hell and damnation than any he or his colleagues had ever produced. In his sermon *Power and Policy of Satan*, Erskine used frightful names for the devil 180 times and referred to hell, according to Robe's count, 27 times. Further, Robe wrote that he had observed one follower of the Seceders enjoying the revels of religious enthusiasm.

> [W]ith great Outcries and Agonies of Body [he] declare[d], as he pretended by immediate Inspiration, That the Work of Cambuslang was all a Delusion; that several People who had been at it were Witches; . . . that a certain person he then nam'd was dead and in Heaven, tho' alive in the Month of November last; and many other Things he declared, as he said, by *immediate Revelation*. . . .

Finally, Robe used the Seceders' own methods in turning to history—the Scottish reformed tradition. He claimed that his revivals followed in the illustrious tradition of Shotts, Stewarton, and the revivals in Ireland. Noting that the Seceders accepted those revivals because they were led by Blair and other heroes of the reform tradition, Robe questioned what prohibited his congregation from being part of that tradition.[58]

John Willison's retort to Fisher was more pointed. "Oh what Pleasure can you have in ascribing the most agreeable Appearances of serious Religion to the Devil, except what is to be found in your society?" Willison recognized the central symbolic role served by the covenant within the Seceder's system and refuted Fisher's accusations on this score, not by counterclaims that the revivalists did support the covenant but by public statement that the era of the covenant was over.

> Because we speak about Church-government in such moderate and scriptural Terms, you raise the Cry against us. . . . Do you think that a Minister that is *Independent* or *Episcopal*, yet strictly *Calvinist*, with evident Characters of his having the Spirit of the Church, and holiness in Heart and Life, is of greater Esteem in the Sight of God; than a *Presbyterian Minister* who may be loose in his Morals, and *Arminian* or worse in his Principles?[59]

Perhaps the most scathing attack, published anonymously ("by a soldier"), accused the Seceders of jealousy and conceit in the face of spiritual success that was not their doing. Poor George Whitefield

> ye extolled to the Skies as a singular Instrument in the Hand of GOD for to convert Souls, but unluckily failing to join with you, he's become an *Enthusiast,* and his Doctrine so sound and orthodox before, alas! is for this become a Delusion.

This pamphlet was written in response to an act of the Associate Presbytery proclaiming a fast day. The author asked, "Had you told your Hearers that the main Cause of this fast was out of Envy and Malice? because great numbers are drawn from your Communion?"[60]

The Seceders and revivalists obviously shared some forms and rituals. Both indulged in sermons of sin and judgment designed to stir the people into a heated, emotional state. The accusations against and defenses of each indicate that they also shared an allowance for popular outpourings. The Seceders channeled these activities along rigidly defined patterns of response that centered upon invocation of the past, specifically the Solemn League and Covenant. The revivalist preachers observed a less formalized, but just as ritualized, pattern of spontaneous responses. These were channeled into the work of individual conversion.

All reports of the individual conversions followed the same, careful schedule. The "awakening" would be to the decrepit, hopeless state of the soul and the punishment that awaited. The bodily contortions and other physical abnormalities generally accompanied this early stage, though they might follow the sinner to the next stage of utter despair in the face of the absolute justice of the penalty. This was followed by an indeterminate period of hope as the candidate discovered Christ, hope that climaxed in the happiness of assurance, the latter often accompanied by tears. After this the convert would fight a constant battle to follow in the way of God as was owed by a privileged, saved being to his creator and savior.[61]

These stages of conversion were much like those ritualized in the communion services. Not surprisingly, the communion service, and not the daily preaching, remained the center of religious life. After the peak of the revival had passed, the session at Cambuslang called for a celebration of the Lord's Supper even though the sacrament had been held only three months before, for "many who had a design to have join'd in partaking of that Ordinance were kept back, thro' inward discouragements or outward Impediments."[62] At Kilsyth also the congregation was inspired to a second communion within three months. Robe invited eleven ministers to assist him, and several more simply arrived and helped. The people, nearly 1,500 communicants, heard sermons in the fields all day of the fast day, in the church Friday evening, and in both field and church all day Saturday. The Sunday distribution lasted twelve hours and included twenty-two separate services. Sunday evening, all day Monday, and Tuesday morning were filled with preaching.[63]

George Whitefield did not visit Ulster, and there were no Irish revivals like those at Cambuslang and Kilsyth. It may well be that without Whitefield such revivalism would not have occurred. The Irish Presbyterians clearly rejected the progressive influence in the synod, but they seemed to find adequate spiritual support and religious outlet among the more evangelical members of the synod and among the Seceders. Futhermore, the importance of the subscription controversy to this development should not be underestimated. Much reforming energy had been used up and diffused by the six-year effort to exclude the nonsubscribers. The ritualization of subscription of the Westminster Confession could, and did for a time, replace the other traditional conversion rituals. The extreme evangelicals remained unsatisfied, but the Secession Church fed those needs in both Scotland and Ireland.

Neither the Ulster Synod nor the General Assembly of Scotland supported the eighteenth-century revivals. In Ireland the synod self-consciously paid obeisance to the laity's fears by endorsing the ritual of subscription as a test of orthodoxy. The ministers also worked very hard to maintain tradition in order to keep the Seceders out, yet the synod was never willing to give the congregations all that had been requested. For this reason the official church, with its authority, was in frequent conflict with its congregations, who had the financial power.

The Scottish assembly was even less sympathetic to the pleas of some congregations, refusing to change its direction for the pleasure of the people. The people had no power, except to leave the pastor's care for a nonconformist; certainly they had no financial control. Congregations across Presbyterian Scotland, but especially the western and central regions, continued to support the Secession Church and the revivalism brought by Whitefield in 1742. The laity also did find an increasing number of ministers to be enthusiastic supporters of their activities. Indeed in 1751 the evangelical movement across Scotland pulled together in the impressive Concert of Prayer, calling for the reform of church and nation. The very nature of this call, however, indicates that while the evangelicals represented a significant portion of the laity and, to a lesser extent, the clergy, the moderate party, the intellectual elite among the clergy, remained in control of the official church. Because of the lack of support among powerful clergymen, revivalism in the eighteenth century remained outside the establishment.

Ironically, in the first half of the nineteenth century, both national churches would experience a major reform and return to the old Calvinist orthodoxy and revivalism, bringing with this return renewed commitment of the laity. However, in 1760 this was far in the future, and what was actually observable was a clergy largely committed to erudition and moral behavior, leading congregations that would regularly lose members to the new traditional movements.

Considering the violent hostility that was nourished between the Whitefield followers and the Seceders, it seems most ahistorical to identify the two groups as offshoots of the same tradition. Yet that was precisely what they were: products of the revered, reformed Protestant ethos. With the appearance of these two disparate systems, the core of popular tradition could be easily distinguished from the similar, unessential features that most manifestations exhibited. Popular, emotional participation had always been a part of this tradition. From the public acceptance of John Knox through the Six-Mile-Water Revival and the suppressed conventicles to the signings of the Solemn League and Covenant, the laity sought active participation in their church's rituals. Ideology was far less important, although basic adherence to the tenets of Calvinism was required. What was firmly at the center of this tradition was the ritualized experience of community conversion. The 1630 revivals, the activities of the Remonstrators, the conventiclers, and now the Seceders all worked toward a pure, renewed community of God. Even the Cambuslang revivalists, enjoying the individual conversion experience prescribed by Whitefield, realized their true goal in the communions. The large communion services enjoyed by 1,500 people still lasted several days. Supposedly involving the recently converted, the communion services were structured with initial days of fasting and prayer, followed by repentance and reconciliation. Through these ritualized forms, the people found themselves able to return their community to God and once again find spiritual peace in their common bond.

5

Puritans, Presbyterians, and the Promise of Community

In 1630, the year of the great communion at Shotts, a large group of English nonconformists emigrated to the new world colony of Massachusetts Bay. Fleeing the confusion and persecution of the 1630s, these Puritans migrated by the hundreds, in families, planning to construct a new society. Within ten years the colonists had scattered southward down the coast and through the Connecticut valley; in another twenty years the Puritans had reached northern East Jersey and Long Island. By the end of the century, the Jersey towns of Cohanzy, Elizabethtown, Newark, Woodbridge, and Freehold and the Long Island towns of East Hampton, South Hampton, Southold, Hempstead, Brookhaven, Jamaica, Newtown, and Flushing all had organized congregations led by ministers formally educated at Harvard, Yale, or a British institution.[1]

The latter half of the seventeenth century also saw the founding of two colonies in the middle-Atlantic region, the Jerseys, founded in 1665 and consolidated into a single colony in 1702, and Pennsylvania, established in 1682. Among the proprietary owners of East and West Jersey were several Scots who encouraged their own countrymen to migrate there. Also among the investors was William Penn, the sole proprietor of Pennsylvania. Both Penn and his fellow proprietors of Jersey eagerly sought colonists of all trades, but particularly farmers, to develop the land and transform the wilderness into a profitable enterprise. They advertised for immigrants throughout Northern Europe, and the fertile land, so cheap and available, was no small incentive. Moreover, committed on principle to religious toleration, the proprietors used their official policy of toleration to appeal to dissenters in various regions who suffered persecution at the hands of state churches.

Thus were the middle colonies prepared to receive dissatisfied immigrants. Economic bounty and religious tolerance must have held an attraction irresistible to any people truly suffering under persecution and poverty. As the invitation was distributed throughout the late seventeenth and into the eighteenth century, thousands of people from the poor and middling ranks managed to cross the Atlantic and settle in the middle colonies. Among the first to take advantage of the opportunities available in this region were the Scots-Irish Presbyterians of northern Ireland.

The Irish migration began slowly in the eighteenth century, becoming a continuous flow by the 1720s. The Irish did try New England first but soon switched their destination to the middle colonies, for New England proved unwelcoming to the Scots-Irish. While the Scots-Irish owned very little material wealth, they did bring their unique culture and religious system. Of course, when they arrived they found a Presbyterian church awaiting them, constructed by the region's Puritan communities along with the resident Scottish and Irish ministers. As the numbers and proportion of Irish Presbyterians increased, the church, initially English in tone, would gradually conform to the Irish religious experience.

Almost immediately after they arrived, the dogmatic Scots-Irish clashed with the infant Presbyterian church dominated by English and New England Calvinists. Within a decade, in the 1720s, the ministers had reached a constitutional compromise. Aroused by the public spectacle of heterodox preachers ordained for the colonies by the Irish church, the people and the ministers united and forgot their regional differences in the face of this threat. This uneasy unanimity would remain for scarcely three years until, in 1738, an entirely new set of problems would realign factions, transcending regional differences, involving the laity whole-heartedly, and splitting violently the church community.

Although both the cultural groups that constituted the middle-colony Presbyterian community should be considered in some depth, this analysis, beginning with the Ulster emigration, will focus upon the Scots-Irish dimensions. The early conflicts between the New Englanders and the Scots-Irish clearly paralleled the problems experienced by the Irish church. In fact, New England and Irish clerics mirrored precisely the behavior of the ministerial communities of southern and northern Ireland, respectively. Moreover, several rituals seem to have been imported directly from Ireland. The religious goals of both the New Englanders and the Scots-Irish were the promotion and maintenance of pure, godly communities. Nevertheless, the methods that the colonial Presbyterian church chose to achieve those goals were carried over from Ireland.

Ireland provided the perfect model for the colonial church. In Scotland and New England the Calvinist communions constituted the established church, whereas in Ireland, as in the middle colonies, the Presbyterians were a dissenting minority. In both regions the Presbyterians lacked the official financial support of tithes and were therefore dependent upon voluntary contributions from the membership. Members of both communities were generally excluded from positions of civil power and influence, either by legal restriction or by economic and cultural circumstances. Finally, since each community existed within a culturally pluralistic society, both had to define their religious systems carefully in order to establish and maintain a cultural identity separate from the surrounding environment. In this primary way the history of the colonial Presbyterian church repeated that of the Irish church: the story of a determination not to conform to an English, civilly protected establishment.

The reign of William III had brought a salve to the wounds of the persecuted Irish community, but the succession of Anne heralded a return to the uneasy, insecure position that dissenters had generally occupied under Stuart power. While the secular government refrained from any prejudicial action, except for the stoppage of the regium donum in 1714, the sacred authorities were granted the means and resources to prosecute dissenters. The Church of Ireland had always been jealous of the large Presbyterian membership in the north, but the bishops had accepted that membership, provided that the Presbyterian influence remained confined to the north. During the early years of the eighteenth century, however, the Anglicans began to fear that the Presbyterian communion was moving south. As the bishop of Kilmore wrote concerning the Presbyterians of Belturbet,

> I think it my duty to join most earnestly with them that if it be possible, this growing faction may be supprest: least in time they should think themselves powerful enough to oppose the Government, in the South as well as in the North.[2]

The bishops always emphasized the threat of rebellion in the hope of inspiring the civil government to greater severity. However, it is far more likely that the bishops feared not political trouble but congregational desertion. Buried in his concerns for royal supremacy, the bishop of Kilmore mentioned that the Anglican minister in Belturbet had left his charge, encouraging the "schismatic" to take over.

Except for the brief support of Anne during the last year of her rule, the Church of Ireland proved unable to mobilize the British government behind its cause. George I, Anne's successor, was a Calvinist himself, and while he seemed happy to join the Anglican church as a condition of his kingship, he had no intention of persecuting other Calvinists. In 1718 the English parliament passed an act of toleration for dissenters, and although

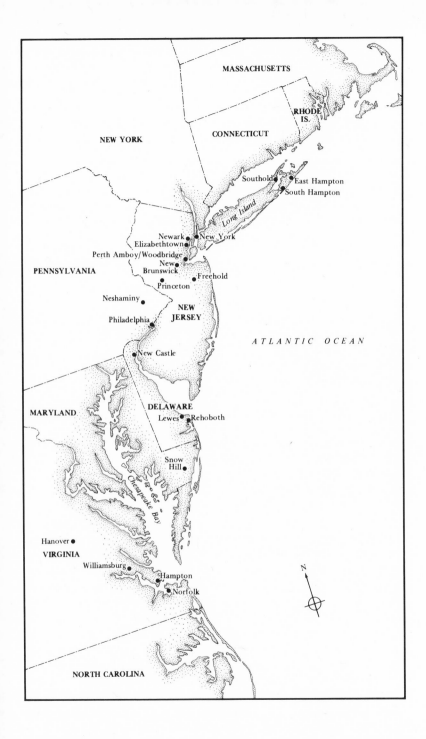

139

staunchly opposed by the Irish bishops, the measure was passed for Ireland as well. Unfortunately, the act only permitted the dissenters to enjoy their religious services unimpaired; it did not erase other penalties imposed for nonconformity. The Presbyterians were still prohibited from holding political office by the Test Act. More important, Church of Ireland courts still had the means to harass dissenters. Bishops retained authority over marriages and licenses, and any couple not married by an episcopally ordained minister was liable to prosecution for immoral conduct. The children of Presbyterian unions could be considered legal bastards; if someone challenged the parental union in probate court, those children would be disallowed as heirs. This situation was not remedied until the nineteenth century.

The Irish bishops notwithstanding, the Irish governors had no desire to offend and alienate the Presbyterian inhabitants. The native Irish far outnumbered the combined English and Scottish Protestant populations, and without the Scots-Irish presence the English feared that they would be unable to control the natives in the north. When the serious emigration began in 1717 the English government worried; when the flow radically increased at the end of the 1720s, the government decided to examine the problem and take action to stop the massive desertion from Ulster.

The Presbyterian ministers continued to decry the activities of the Anglican authorities against them, but other documents indicate that the primary problems were economic. During the early decades of the eighteenth century many leases granted during the plantation years came up for renegotiation, and tenants had to settle for far less favorable terms. In a letter to Robert Craghead written in 1729, Francis Iredell did blame the proceedings against Presbyterian marriages and the efforts to close schools taught by nonconforming schoolmasters. However, the primary reason for emigration, he claimed, was poverty. For the previous three years the harvests had been exceptionally poor, and the difficulties that the crop failures brought were exacerbated by landlords' raising the rents beyond the tenants' ability to pay. Those whose rents had remained reasonable feared that this would soon change. Added to this were extortionist abuses of the tithe system that went uncorrected by unsympathetic, corrupt judges of assizes.[3]

When John St. Leger and Michael Ward compiled a report for the Lord Justices in June 1729, they also emphasized the economic disabilities. Harvests were poor, and the cost of corn high, while the price received for linen yarn had dropped to half. Rents were outrageous, often higher than the land's productive value, for the fees were frequently set for an absentee landlord by an agent seeking a higher percentage profit. Some landholders had turned out the Protestant tenants in favor of natives who would agree to a higher rent. Also, in order to ensure ability to raise rents the landlords gave very short leases, and this discouraged improvements. The longest lease allowed on church lands, for example, was twenty-one years, and most leases did not last beyond seven. The Church of Ireland was to blame for setting tithes higher than farm values justified, and for summoning to bishop's court those who could not afford to pay. There was, the report noted, even a tithe on potatoes, which the people considered "harder inasmuch as by cultivating barren heathy land for potatoes they bring it to be good corn land for ever after." The Test Act was also discussed, since so many Presbyterian spokesmen had mentioned it, but the authors had heard of no prosecutions. As if to underline the hardships of the people and the strength of their discontent, the report listed the claims by which captains and shipowners attracted their passengers to America—much land, little or no rent, no taxes, and no tithes.[4] Even those who discounted the idyllic claims made for the New World admitted that some tenants had proper grievances against landlords, "tythmongers," and some of the county courts and justices. The people were responding to reports from the colonies that "all men are there upon a levell and that it is a good poor mans Country, where there are noe oppressions of any kind whatsoever."

> [Y]e Richer Part Say, that if they Stay in Ireland, their Children will be Slaves, and that it is better for them to make money of their Leases while they are worth Something to inable them to transport themselves and families to America, a place where they are Sure of better treatment, although they shall meet with some hardships, they are very well assured their posterity will be forever happy.[5]

Fed up with economic woes and official troubles, people throughout northern Ireland had access to ships to the New World through five different ports—Larne, Derry, Coleraine, Belfast, and Sligo. Some passengers were persons of comfortable means, able to pay for their passage; these hoped to use their limited resources to greater advantage in the colonies. Others were actually among the wealthiest segment, including English and Scots, who expected to achieve positions of influence by virtue of their superior holdings. The overwhelming majority, however, were of the poorer sort, willingly obliging themselves as indentured servants in the knowledge that the Ulster situation was desperate, while America held hope. In the words of one promoter, "There is Servants come here out of Ereland, and have serv'd their Time here, who are now Justices of the Piece [*sic*]. . . ."[6]

During the sixty years preceding the American Revolution, thousands of Irish Protestants migrated to the colonies. Finding the New England

Calvinists inhospitable to the Irish style of church worship and government, the migrants turned their attention southward. A few went to the
Chesapeake; a significant segment, to South Carolina; but the largest number, by far, settled in the middle colonies, arriving through the port of
Philadelphia. The reports of the overwhelmed chroniclers of the period
indicate that the total number of migrants was upwards of 500,000. Even
considering gross exaggeration by the reporters, the shipping and port
notices in contemporary newspapers and the headcounts performed at the
request of an anxious Irish government demonstrated a massive migration
of well over 150,000–200,000.[7]

The Irish government first grew concerned in 1728 because of rumors
of massive migration. One observer had estimated that between 1720 and
1728 approximately 40,000 people had left the country; he himself had
counted seventeen ships departing in 1728 alone. A second report from
Lisburn claimed that 10,000 people had already left the area, with twenty
ships standing ready to take on passengers. During that same summer
Archbishop Boulter recorded the exodus of 3,100 individuals, 10 percent of
whom could pay their own passage. In response to such frightening
reports, the civil authority finally requested a formal count for the previous
three years. Those official totals stated that 4,062 persons had left through
the various northern ports in the previous three years, plus a few from
Belfast going to Virginia, and another two hundred who had sailed from
Duadilk. From Derry alone, merchant Robert Gambie reported in 1729 that
during that summer twenty-five ships were expected to carry at least 140
passengers each to America.[8]

The numbers only increased as the years passed. The *Pennsylvania
Gazette* reported that in September 1736 one thousand families left Belfast;
one hundred persons had arrived in Philadelphia on the ninth, a similar
number had debarked at New Castle, and twenty-three ships were expected
daily. Arthur Young, in his *Tour of Ireland*, estimated that between 1740
and 1760, the ships leaving Derry carried 2,400 passengers annually.
Scholars have estimated that between 1730 and 1769, over 70,000 Ulster
Presbyterians arrived in the New World. No wonder that in 1744 the
Presbyterian minister Samuel Blair could write, "All our congregations in
Pennsylvania except two or three chiefly are made up of people from
Ireland."[9]

When the Irish immigrants began to arrive en masse in Philadelphia in
the late 1720s, they found a congregation and a Presbyterian synod awaiting them. The Presbyterian congregation in Philadelphia had been organized in 1697 as the only Calvinist alternative to the Quaker meeting and
the Anglican church. The congregation, composed primarily of English

people and New Englanders, in 1701 called Jedidiah Andrews, a Harvard graduate of 1695, as their first minister. The patronage of a few prominent citizens, notably William Allen, provided this church with an early stability; and while new Presbyterian arrivals usually moved quickly to acquire land and establish small congregations in the countryside, this older, stable Philadelphia community early became the political center of the newly formed Presbyterian network.

The first presbytery in the North American colonies, the Presbytery of Philadelphia, was formed in 1706. Francis Makemie, pastor of a church built on his own land in Rehoboth, Maryland, seems to have been the driving force behind the organization.[10] He had previously brought to Maryland from his home presbytery of Laggan, Ireland, John Hampton (preaching at Snow Hill) and from Scotland George McNish (preaching at Monokim and Wicomico). Also in Maryland was Nathaniel Taylor, of Scotland, preaching at Patuxent. Makemie encouraged these three Maryland clerics to unite with New Englander John Wilson of New Castle (Delaware), Irishman Samuel Davies of Lewes (Delaware), and Andrews.[11]

Makemie may have chosen Philadelphia as headquarters because it was the farthest south of the major colonial cities within those areas that contained known Presbyterian communities and the closest city to his Maryland community, or because of its importance as a port of arrival. He may also have been attracted by its central location, from which he could call and organize many more nonconformist ministers and congregations, most of whom (excluding dissenters in New England who were already organized) resided in New York, New Jersey, and Pennsylvania. Of some importance must have been Pennsylvania's policy of religious toleration that allowed the presbytery to meet in peace. Makemie hoped that this Presbyterian organization would serve the middle colonies as an administrative support for dissenting ministers, especially in the face of a colonial Anglican community growing in numbers, strength, and authority. As Makemie explained to Benjamin Coleman of Boston:

> Our design is to meet yearly, and oftener, if necessary, to consult the most proper measures for advancing religion, and propagating Christianity in our Various Stations, and to maintain such a correspondence as may conduce to the improvement of our Ministeriall ability by prescribing Texts to be preached on by two of our number at every meeting, which performance is Subjected to the censure of our Brethren. . . .[12]

Despite Makemie's death in 1708, the presbytery grew quickly, acquiring ten ministers in as many years. In 1716 the presbytery decided that the size of the growing community, combined with its wide geographic distribution, justified a division into three presbyteries under the headship of a new Synod of Philadelphia. The presbyteries of Philadelphia, Long Island, and New Castle flourished so that in 1730, only twenty-five years after the church's initial organizing efforts, the Synod of Philadelphia had thirty ministers.[13] Despite the rapid expansion of Presbyterianism in the middle colonies, the Philadelphia congregation remained small, containable, as of 1730, in a single church edifice that held perhaps a hundred families. The migrants did not stay in the city but tended rather to spread themselves out, setting up an early pattern of small, scattered congregations. During these early years, because of the scarcity of ministers and the inability of small congregations to pay reasonable salaries, a minister would often have to serve three or four congregations spread over as many as fifty square miles.

The structure of the original presbytery and the authority granted to the presbytery by its members have continued to puzzle church historians. Was it an official assembly with power to bind its members, as in Scotland or Ireland? Or was it more like the consociations of Connecticut and western Massachusetts, voluntary associations of ministers who might advise each other but without authority to interfere? Certainly Makemie was a Presbyterian, educated at the University of Glasgow and ordained by the Presbytery of Laggan, Synod of Ulster. Hampton and Davies were also from Ireland, and McNish and Taylor from Scotland, undoubtedly of Scottish Presbyterian persuasion and training. Furthermore, some of the functions performed by this union were "presbyterian." The presbytery supervised congregations not under an individual pastor's care, assigning various members to preach in vacant pulpits throughout the year. It examined candidates and probationers, licensed preachers, and ordained ministers over congregations. In one instance the body intervened in a dispute between one of its members and his congregation, in the end finding fault with both parties and dissolving the pastoral relation.

Ironically, this particular case also demonstrates the tenuous authority exercised by the group as a judicature. The people at Woodbridge, New Jersey, had asked the presbytery to interfere in their battle with their pastor, Nathaniel Wade. Since Wade had been a member of a New England consociation, the presbytery felt that it could only play a moderate, mediative role. A fight had been flourishing within the congregation that seemed to represent the opposing views of Congregationalists and Presbyterians, actually between the residents of Woodbridge, originally from Newbury, Massachusetts, and those of Perth Amboy, recently arrived from Scotland. The arguments involved the ability of Wade to serve the people of Woodbridge, especially since he had joined the presbytery; his chief opponents were New Englanders. The peculiar circumstances surrounding the dissension forced the presbytery to consult Boston, and only through the influence of Increase Mather was Wade convinced to demit his pastorate and the congregation persuaded to agree upon a call to another minister.[14]

This minor incident reveals a tension felt throughout the formative years between the structured jurisdictional system carried over by the Scottish and Irish Presbyterians and the more volatile clerical associations of New England. Not all the New England Puritans were Congregationalists. Coming from an atmosphere of English dissent where nonconformity was defined as opposition to the Church of England, the Puritan immigrants did disagree over questions of polity, church membership, and the ministry—disagreements which could only surface in the absence of any common religious opposition.

The Congregational party dominated Massachusetts, while Presbyterian sympathizers in New England either accepted the decision and lived as the majority prescribed or else fled beyond Massachusetts' control. However, the communities of Connecticut and therefore Long Island and New Jersey were largely English Presbyterian. This identification does not in any way imply that these New Englanders were in complete sympathy with the Scots, for the Puritans refused to accept the primacy of the Church of Scotland. They did share many common views of theology and church structure, but not all, and they seemed to characterize themselves as being in opposition to the Congregationalists. In fact, a survey of the characteristics that distinguished this group from other English dissenters shows that, more than the Scottish church or even the Ulster Synod, the transplanted New England communities resembled the Southern Association centered at Dublin.

First, the English Presbyterians, like their Scottish counterparts, were committed to an educated ministry, a commitment not shared by all the nonconformist congregations in England. The fitness of a minister was, therefore, determined by other ministers and not by the congregation; further, a minister was ordained by other ministers and not by the congregation that called him. That is, the actual ritual of the laying on of hands was performed by clerics, not laymen. Some dissenting clergymen went so far as to ordain ministers even though the candidates had received no clear pastoral charge, an institutionalization of the profession beyond even the Scottish structure. In the Scottish and Irish churches, only ministers determined to work as missionaries could be ordained *sine titulo*, that is, without a charge.

Second, the process by which candidates for ordination were approved required a committee of ministers, and for this reason and others many New England ministers encouraged other clergy to join them in voluntary associations. These associations were primarily educative, and membership was never mandatory. The only officially sanctioned duty of such groups was the approval of ministerial candidates and their ordination; yet even these functions were regularly performed outside any organized network. Unlike Scottish presbyteries, these associations had no lay members and thus had absolutely no authority over congregations or individual churches. Although such a consociation could be asked to give advice in the case of a cleric accused of immoral conduct or heretical beliefs, the opinion of that body was never binding. This was far removed from the Scottish Presbyterian system which consisted of a hierarchical series of legislative/judicial units—congregation, presbytery, synod, national assembly— each having authority over all clerics, lay persons, and subordinate units within its jurisdiction.

A third characteristic is clearly represented in the behavior of the Southern Association of Dublin during the 1720s. English Presbyterians reacted strongly against any sort of subscription to a creed or written constitution, doubtless in response to the requirements of the Anglican church that demanded open subscription to several statements of beliefs and loyalties. While these ministers, like all good Calvinists, certainly believed in the possibility of heresy, they also believed that within such an educated community as their own, the method prescribed by the New Testament for accusations and trials of heresy was a far better preventative than any rules requiring subscription could be. Moreover, these Presbyterians suspected that subscription requirements could pave the way to heresy, since a false prophet could hide behind a signature on a document. This attitude toward subscription differed radically from the posture of the Scots and the Ulster Irish who variously embraced and demanded or took for granted credal subscription as part of religious identity.

These differing beliefs regarding church government and polity might have blocked any ministerial union between these English dissenters and Scottish and Irish Presbyterians were it not for a shared concept of church membership, a similar, Calvinist understanding of salvation, and a common sacramental theology. Presbyterians rejected the possibility of a visible church. English Independents and many of their followers in New England were confident that one could tell with a fair degree of certainty who was saved and who was condemned, and they believed that it was proper to restrict church membership to those individuals who were saved. English Presbyterians followed the more liberal notion that since it was impossible to tell truly who was and was not saved, the best course would be to allow everyone to join the congregation except, of course, flagrant, notorious sinners whose participation in sacramental rites would be an affront to God and the community. A closed membership gave the church members a sense of spiritual righteousness not available to the Presbyterians. On the other hand, a parish system expanded the control of a church, its minister, and its governing body over the entire town.

Following directly from these varying perceptions of visible church versus parish church were different conceptions of the sacraments. Since the Independent congregations limited their membership to those who could provide proof of conversion, an experience easily identified within a specific ritualized structure, anyone who had not experienced conversion was refused admittance to the sacraments. (Within a theology that accepted infant baptism, this restriction was applied through the status of the parents of infant baptismal candidates.) Thus the sacraments were seen as sealing ordinances, signs of divine grace already received, the sacred communication between God and his elect. As the Presbyterians opened their membership rolls to all but a few gross sinners, participation in the sacraments was opened. Presbyterians did believe that individuals experienced conversion and that such conversion was necessary to spiritual salvation, but they also recognized that saving grace did not follow the same path in everyone. Therefore, they avoided judging a person's spiritual status and instead concentrated efforts upon allowing grace to do its saving work. Within this philosophy the sacraments, especially the Lord's Supper, were tools through which the Holy Spirit worked to effect salvation. "The Gospel preached, and its Ordinances administered, have always influence upon the Souls of the hearers or receivers, bringing them nearer to Salvation or Damnation. . . ." This was not due to any miraculous quality of the sermons or sacraments themselves, but to the foreordination of God. "So it is God and his Power that gives the effect, and the outward means is to be used only because of his Appointments and Institutions, because it is his Pleasure."[15]

These common beliefs plus, undoubtedly, a sense of being lone Calvinists among Anglicans and Quakers allowed the ministers on Long Island to join with their predominantly Scots-Irish colleagues. When the Presbytery of Philadelphia decided in 1716 to divide into four under a single synod, the leaders recommended that George McNish and Samuel Pumry invite the other clerics in their neighborhood to attach themselves to the Presbytery of Long Island. The very first synod meeting of 1717 was attended by Jonathan Dickinson, the bright young pastor of Elizabethtown, and John Pierson, the reconciling pastor of Woodbridge. With the addition of these two leading New Jersey clergymen the Long Island Presbytery was successfully convened, and for five years the judicatures administered their affairs peacefully and without contention.

Controversy in the clerical ranks was first evident in 1721 when George Gillespie, ordained seven years before by the Presbytery of Glasgow, presented the following overture to the synod:

[I]f any Brother have any Overture to offer to be formed into an Act by the Synod for the better carrying on in the Matters of our Government and Discipline, yt he may bring it against the next Synod.

The resolution seemed rather vague and innocuous, but battle was engaged over the words "Overture," "Act," and "Government and Discipline." Up to this time the presbyteries and synod had restricted their activities to the administrative functions of testing ministerial candidates, supplying vacant congregations that wanted preaching or the distribution of the sacraments, intervening, when invited, in disputes between pastors and their charges, and supervising the behavior and theology of ministers. This overture suggested that the synod would have authority to legislate as well as to execute church laws. The overture was passed, but not without the open protest of six members, a large proportion when the total membership numbered only twenty-five.[16]

Why Gillespie brought forward this resolution is unknown. Perhaps he was merely trying to clarify procedures in the developing church. A more likely possibility was that Gillespie had an overture that he himself wanted to present. He had some definite, censorious opinions concerning the lax discipline exercised by the new synod and, in fact, had recently protested the mild sentence (four-week suspension from the ministry that would be lifted without trial at the end of the four weeks) levied on Robert Cross, a minister found guilty of fornication.[17] Probably Gillespie had no intention of forcing the American church in a Scottish direction and alienating the New England contingent. Yet the six protesting ministers, led by Jonathan Dickinson, included four New Englanders and two Welshmen.

Jedidiah Andrews quickly set about healing the division. In a letter to Benjamin Coleman of Boston in 1722, Andrews declared, "The difference is in words, for I can't find any real difference, having sifted the matter in several letters which have passed between Mr. Dickson and me upon it." Andrews acknowledged the excellent compromise in the Pacific Acts passed by the Irish church in 1720, demonstrating an awareness of the controversy in Ulster. He ascribed the protest to "the squabble at New York" between the Puritan residents and the newly arrived Scots-Irish. If the ministers were sensing an early ethnic conflict, it would explain their immediate hostility and determination to oppose the measure.[18]

Jonathan Dickinson, as moderator of the 1721 Synod, opened the Synod of 1722 with a sermon based upon Paul's second epistle to Timothy; the subject was scriptural integrity and church government. Beginning with an assertion "That the *Holy Scriptures* are every way sufficient to make him [the minister] perfect in, and throughly [*sic*] him for the Whole Work of his Ministry," Dickinson declared the scriptures self-sufficient. The whole path of salvation was there; the Sermon on the Mount should provide the law of discipline:

> [T]o Institute any new PART of worship, or to bring any thing into Gods immediate Service, not expressly instituted by Christ, is a bold Invasion of his Royalty, who is *Head over all Things* to his Church.

Thus Dickinson saw the imposition of any act or constitution as absolutely unjustifiable. He did affirm that the Presbyterian system was the form of church government that best conformed to the Bible, and he admitted the necessity of man-made codes establishing some items—times and places of meetings, rules of order—because the Bible had provided only the general terms. Nevertheless, "Substantials of Government are left upon Record in the Word of God, and are unalterable by any Humane Authority." With simple logic, Dickinson argued that acts paralleling scriptural rules were unnecessary. As to those laws not found in scripture, if the laws were not binding, they were "useless except as bones of contention"; if binding, the laws were a usurpation of Christ's power.[19]

The most amazing aspect of this homiletic response is that this sermon with its well-defined, precise agenda was delivered in answer to an extremely vague overture about the reading of proposed acts one year before consideration. Yet by the end of his sermon Dickinson had gotten to subscription. And while the New Castle Presbytery did require its candidates to subscribe the Westminster Confession, the synod had made little progress in that direction. Nevertheless, Dickinson spoke of creeds and confessions, their usefulness and liabilities. His statements mirrored exactly those of the nonsubscribers in Ireland: ministers have full commission to study and interpret God's laws, "But these haveing no claim to Infallibility, can have no Authority to impose their Interpretation."[20]

During the 1722 Synod, four of the protesting ministers introduced a series of four articles meant to open the way for compromise. In the first statement they accepted the sole authority of the church to legislate aspects of church government, without taking a stand upon the extent of the legal obligation mandated by that legislation. The second allowed that ecclesiastical judicatures should determine "Time, Place and Mode" of carrying on church government, making certain that conditions, methods, and decisions conformed to the biblical injunctions and Calvinist system, following the qualifications in Dickinson's sermon. Here, too, the authors admitted that such decisions could be called "ACTS" and they "will [have] taken no offense at the word, provided yt these Acts be not imposed upon such as conscientiously dissent from them." This clause protected the privilege of any minority within the church community to refuse to adhere to nonessential acts, which would include most actions relating to church government, to which they conscientiously objected. The third and fourth articles stipulated the power of the synod to compose directories on all parts of discipline and establish the principle of hierarchical judicatures, but only under the same conditions as those stated in the second article. With the general acceptance of the judicature system and the open concessions to conscience, the synod members were "so universally pleased with the above said Composure of their Difference that they unanimously joyned together in a Thanksgiving Prayer, and joyful singing of ye 133 Psalm."[21] The unanimous joy lasted until 1727.

In that year Dickinson's fears were realized: the issue of subscription reached the colonies. During the Synod of 1727 the Irish minister John Thomson of New Castle Presbytery moved that the synod as a whole subscribe the Westminster Confession. Jedidiah Andrews wrote that the overture was "not then read in Synod. Means were then used to stave it off, and I was in hopes we should have heard no more of it."[22] However the issue arose again in 1728 when it was proposed that all ministers subscribe or be disowned. The question was again deferred until 1729 when the full synod would meet and each member would be asked to give an opinion. The importance and delicacy of this overture was underlined by the calling of a full synod. In 1724 the synod had validated the representative synod, by which the membership would be proportionately represented by lay and clerical delegates selected within the individual presbyteries. In this case, however, representation might not have been reliable; each minister and elder wanted to speak for himself.[23]

Andrews deeply regretted the surfacing of this issue. He knew that the measure would be carried by numbers, but his own "Countreymen," while willing to accept it as the creed of the church, were not willing to accept the Westminster Confession as a test of orthodoxy. Andrews's major concern with the measure was its divisiveness. Up to this time "the different Countreymen seem to be most delighted in one another," but they "do best when they are by themselves." In other words, as long as individual ministers focused upon local congregational affairs, they worked together splendidly. However, when they began to concentrate on the church as a whole, the ministers would form ethnic alliances from which they would argue their different positions. He wondered briefly if the real goal was to oust the New Englanders from the middle-colony church, especially since "I think all the Scots are on one side, and all the English and Welsh on the other, to a man."[24]

As a justification for his resolution, Thomson published his overture, preceded by thirty pages of explanation. The pamphlet revealed not only his intellectually constructed reasons but also a few hidden motives. Thomson's argument would easily have fit into the prosubscription literature of the Irish controversy. If everyone acknowledged the Westminster Confession as orthodox, he asked, why would they not subscribe it? Since the confession was agreeable to the Word of God, subscription could not be an imposition. The refusal to see his opponents' point of conscientious scruples was, again as in Ireland, answered by the inability of subscribers to join in communion with those who did not subscribe. The primary focus of his anxiety was the prevalence of heresy. Thomson rooted his debate in the duty of all Christians, especially ministers, and especially the church, to defend the gospel. Moreover, the church would perpetuate for posterity the truth uncorrupted; therefore the institution needed to be fortified against an invasion against truth. After several paragraphs more upon the duty of the church and the general weakness of the times, Thomson waved the banner of subscription as a standard by which error might be kept out of the church. Such a standard was required to counterbalance the fact that so many ministers

> have the Edge of their Zeal against prevailing Errors of the time very much blunted, partly by their being dispirited, and so by a kind of Cowardice are afraid, boldly, openly, and zealously to appear against those Errors that show themselves in the World under the Patronage and Protection of so many Persons of Note and Figure; partly by a kind of an Indifference, and mistaken Charity, whereby they think that they ought to bear with others, tho' differing from them in Opinion about points that are Mysterious and sublime, but not practical nor Fundamental, such as Predestination.[25]

The agenda hidden beneath the polemic against heresy had much more to do with the conflict as analyzed by Andrews. The pamphlet was titled an overture to the "Synod of dissenting Ministers," not "Presbyterian ministers," and several times Thomson indicated that this was an issue. He cited the Church of Scotland as an example to be followed; he called for subscription as a necessary precursor for union with that body. Yet several

pages later he denied any national bias. He argued that his position on
subscription was not the result of his nationality, but the result of sound
logic.[26] Thomson probably wanted to be objective; he may have thought
himself objective. Nevertheless, his cultural identity as Scots-Irish was
well demonstrated by his apparent willingness to split with the New
Englanders.

> I acknowledge Peace and Unity to be sweet and beautiful, when joined with
> Truth and a good Conscience; but when they cannot be joined together, I am
> sure Truth and a good Conscience is infinitely preferable to any Peace or
> Unity that can be had without it.

He remained as willing as his Irish counterparts to suffer schism for the
sake of truth. And, like his models, he was willing to use the laity to
catalyze a crisis. As he justified his pamphlet in the introduction:

> I conceive our Flocks or Hearers they and their Posterity, are no less, yet I
> have almost said, more concerned in the Issue that it shall have in the Synod,
> than the Ministers themselves personally considered. . . .

In other words, he published the overture, complete with lengthy debate, in
order to "inform" a laity that he knew "had so great Interest in all our
Ministerial Acts."[27]

Once again, Dickinson became the spokesman for the nonsubscribing
party, publishing his own pamphlet a few months after Thomson's was
available. Just as Thomson covered the argument of the Irish subscribers,
Dickinson almost point by point reiterated the position of the nonsubscribers in Ireland. Subscription was not necessary or useful. Any hypocrite
could easily subscribe a creed, so this requirement would not discriminate
hypocritical from sincere Christians. Furthermore, while he admitted that
various individuals interpreted the scriptures differently to suit their own
goals, he wondered whether it would not be even easier to perform the
same kind of interpretation with creeds and confessions.

> And thus if *Subscription* should universally obtain among us, everyone
> would subscribe in his own Sence, and all severally put their various, and it
> may be contrary Constructions upon the same Article of our *Confession*.

He then demonstrated that the article on justification could be interpreted in
three separate ways to suit three different theological systems—anti-
nominan, Arminian, and correct Calvinist interpretation. If Arminians and
even Arians could accept the Westminster Confession as the confession of
their own faith, what kind of safeguard was it?[28]

Dickinson favored the presbyterial method of discerning and controlling heresy, the method put forward in the confession as dictated by

scripture. The use of sequential admonitions and eventually trial by prescribed rules protected the church while ensuring fairness to the individual.
The synod also followed established procedures for examining ministerial
candidates and endorsed strict discipline for the clergy, so there should
have been no anxiety about the infiltration of heresy into the Presbyterian
system. Dickinson could express such prescriptions so calmly because he
did not see heresy within the church as a primary problem. From his
vantage point the proximity of heretical denominations like the Church of
England or the Baptists was far more dangerous.[29] Beyond this confidence
that heresy did not threaten the church, Dickinson had a wider tolerance for
and acceptance of beliefs. He accepted a distinction between essential and
inessential articles of faith and was willing to overlook differences of the
latter sort.

> You object that an Erastian, who supposes that Christ has left us at Liberty,
> to chuse what Form of Government we think most agreeable, may freely join
> with us in the actual Exercise of Presbyterian Government. To which I
> answer Why not? I hope your charity is hardly so restricted . . . as to suppose
> this a *damning Heresie* or an *Error* that incapacitates for Communion. . . . I
> can't see what Prejudice these can be to our Constitution, so long as they
> heartily join in our Methods of Discipline; and thereby Declare 'em best for
> us, whether they believe 'em of *Divine Right* or not.[30]

For Dickinson the most important consideration was peaceful union.
Because he worked within an area surrounded by communities of critically
different theologies, he saw a need for the Calvinist churches to join
together, almost as a defense against the majority. Thus he began his
pamphlet with fears of contention; he too referred to Ireland, but negatively: "the Fire of *Subscription* consumed their Glory; and this Engine of
Division broke them to pieces. . . ." Ten pages later he endorsed all the
bonds shared by the united groups, and toward the end he again devoted
several paragraphs to the divisive force of required subscription. He, like
Thomson, displayed his regional bias by citing his own New England as
the model. "The Churches of New England have all continued from their
first foundation nonsubscribers; and yet retain their first Faith and Love."[31]
    Finally, Dickinson did refer to the laity, but only once, and that
indirectly. Nowhere in this discourse did he pretend to answer any request
or objection raised by a lay person, indicating that he believed subscription
concerned only the clergy. His entire discussion of heresy, disunion, and
theological interpretation referred to clerical activities, not lay involvement. In fact, the only time he mentioned the laity was at the very beginning when setting up an initial objection to subscription. If subscription
was to be useful in the maintenance of purity, then congregations as well as

pastors should subscribe. However, such a measure would bring "Schism, Contentions, and Confusions; and even the total Subversion of our Congregations." Thus, concluded Dickinson, no one wanted to do this, although congregational subscription was the logical end of the prosubscription reasoning.[32] What Dickinson failed to grasp was that far from rejecting congregational subscription, the subscribers perceived this as the ideal to be sought. Communicants would be required to affirm the Westminster Confession to receive; parents would have to subscribe on behalf of their infants in order to procure baptism. Dickinson believed it would anger the people, while Thomson perceived subscription as the laity's desire. Both could easily have been correct since they served ethnically different congregations—transplanted New Englanders in Elizabethtown as opposed to Scots-Irish in New Castle.

Dickinson demonstrated that unlike the Ulster nonsubscribers, though very like the Southern Association, he sought unity as his uppermost goal. He and his faction accepted the decision of the 1729 Synod. Of course the subscribers did make concessions to the factional scruples; that is, the synod did discriminate essential from inessential articles of faith, and subscription was required only to the former. "[T]he Synod do not claim or pretend to any Authority of imposing our faith upon other men's Consciences, but do profess our just Dissatisfaction with and Abhorence of such Impositions, and do utterly disclaim all Legislative Power and Authority in the Church." Nevertheless, they believed in keeping the faith "pure and uncorrupt" so that it might pass unscathed to future generations. Therefore, all ministers were required, upon pain of disownment, to subscribe the Westminster Confession and the Longer and Shorter Catechisms as, in essential articles of faith, good expressions of sound doctrine. Any minister who had conscientious scruples with any article was to express them to the judicature, and if the synod decided that the dissent concerned an article not necessary to salvation, the minister might still be admitted to the communion: "none of us will traduce or use any approbrious Terms of those yt differ from us in these extra essential and not necessary points of Doctrine."[33]

Ironically, this measure closely resembled the subscription resolution that had been passed in Ireland in 1705 and in 1720, before the nonsubscribers began their refusals to sign in response to the laity's demand for subscription. Up to that point, presbyteries had peacefully allowed objections to nonessential clauses, and there was no record of anyone's having been denied licensure or ordination on that basis. So too among the members of the colonial synod, the ministers could agree upon which articles were essential and which were not. Every minister previously on the rolls was accepted as a member under the new terms, including many who had several objections each.

Andrews was correct in ascribing this controversy to cultural differences. The original six protesters were either New Englanders or Welshmen, and the active proponents for subscription were either Scots or Irishmen. This observation, however, should not disguise the fact that many clerics, just as in Ireland, did not have strong opinions on the matter. At least one New Englander, Andrews, though clearly opposed to required subscription, devoted all his efforts to reconciling the extreme factions. More important, a large number of Irish ministers remained unconcerned with this problem throughout. These ministers, notably William Tennent and his four sons, did not involve themselves in the dispute. Since the Tennent group, especially the eldest son Gilbert, would soon prove quick to publish their opinions during conflicts, it is telling that over this issue they did not. Perhaps they felt a conflict of identity: the Tennents were well-acquainted with the New Englanders. William Tennent had spent some time in Connecticut, and Gilbert had attended Yale, receiving a master's degree. They may not have wanted to alienate their colleagues. On the other hand, the Tennents were educated within the Irish Protestant network. William senior had been a priest in the Church of Ireland in Antrim. When he arrived in the colonies he left his Anglicanism for Presbyterianism, and the Presbyterianism he had known in Antrim would have been that of Ulster. His wife was the daughter of Gilbert Kennedy, the Irish subscription proponent. The Tennents could have easily accepted subscription as part of the ecclesiastical polity. In fact, the Tennents did all subscribe the confession without qualm or comment.

Two primary factors differentiated this controversy from the battle in Ireland. First, in Ireland the final result was schism; the extremists at both ends were unwilling to accommodate themselves to the ideological needs of the others. In the colonies, however, the moderate reconcilers dominated the assemblage, not only formulating a reasonable compromise position, but convincing opposing advocates to accept that position. The second factor, and perhaps the major cause of the first, involved the participation of the laity. In Ireland, Samuel Haliday had been vindicated, and the Pacific Acts had laid out terms for peaceful coexistence. The laity had again pressed charges against Haliday the very next year. The laity had called for a voluntary subscription. Prominent elders had initiated the divisive process of establishing competing congregations. Many Irish clergymen had said that they had subscribed only in acquiescence to their congregants. In the middle colonies, on the other hand, while Thomson and Dickinson both claimed lay concern, there was, in fact, no evidence of lay involvement. Elders were present at the 1729 Synod, and, presumably, they voted on the resolution. Nevertheless, at no point did the minutes record a lay spokesman for either side, and when the actual subscription was noted, only the ministers' names were listed. Whether or not the elders were asked to subscribe, only the decision of the ministers was considered important enough to be written down. Seemingly the conflict in the colonies represented a difference among the clergy, with factions seeking control of the synod. The ministers, unhampered by lay involvement, were able to achieve a reconciliation and maintain the unity so important to the clergy.

Why did the laity not become involved in the colonial controversy? An easy answer is that the church was new, the Irish colonists were very recent arrivals, and none had time to get involved in esoteric, synodical disputes. This dispute, however, was not necessarily esoteric, as in Ireland this same, general type of congregant was avidly concerned with the issues. Furthermore, presbytery records demonstrate that as early as 1734 the laity did get involved in such discussions. Perhaps the answer lies in the fact that during the 1720s the laity felt no threat to their religiosity. They did not need a protection for their religion because, as far as they could tell, ministers were orthodox and observant of tradition. This conclusion is further strengthened by the behavior of the laity in the 1730s. In two exemplary cases the protest of the laity against unsound ministers set in motion administrative corrections and trials, and in each case the final decision resulted in the removal of that minister from the Presbyterian clergy and the colonial community.

William Orr, a student from Ireland, was licensed by the New Castle Presbytery in 1730.[34] He was ordained pastor of the congregation at Nottingham in 1732 and, when Donegal Presbytery was formed out of New Castle, he was assigned to the new jurisdiction. In April 1734 several elders brought charges against Orr, challenging his orthodoxy. He was accused of preaching against election, employing the standard Arminian criticism. Orr had responded with accusations of slander, and the trial went to presbytery. The presbytery condemned the doctrine that Orr supposedly preached but acquitted him of believing false doctrine, although Orr was warned that he had used questionable expressions. The three elders, John Kirkpatrick, Hugh Kirkpatrick, and John Moor, were rebuked for haste and falsehood and removed from the Nottingham Session. The three acquiesced in the judgment regarding themselves but appealed the judgment regarding Orr's orthodoxy to the synod, who in turn assigned a commission to investigate. The commission worked out a set of nine articles of accommodation among Orr, the congregation, and the complainants, and Orr was again acquitted of heterodoxy.[35]

In June of the very next year Orr requested a demission from Nottingham, as he had experienced severe "disappointments" there. Orr not only decried the considerable arrears, but he accused several congregants of challenging his ministry. Since the demission of a pastorate was always serious, one part of the congregation, represented by Hugh Berry, opposed the demission, and Orr's request included accusations against some of his congregants, the presbytery opted to hold a meeting in Nottingham to investigate.[36] The following September the case was heard, complete with accusations and countercomplaints. Orr again spoke against one congregational faction, led by John Kirkpatrick, accusing them of slander; Kirkpatrick defended his actions as no slander but dissemination of the truth. The proceedings were enlivened by Kirkpatrick's refusal to appear before the presbytery, claiming that his life was in danger, until Berry engaged to guarantee the safety of Kirkpatrick and his followers. The discussion of the demission and Orr's orthodoxy quickly degenerated into a trial of Orr's moral conduct, complete with general, unsubstantiated gossip. After three days of contradictory testimony, the presbytery granted a demission on the grounds that neither Orr, the people, the session, nor the presbytery saw any hope of his further usefulness there. The presbytery again acquitted Orr of the charges brought and issued a series of rebukes to various participants for unbecoming conduct before the judicature.[37]

Still the matter did not rest. John Kirkpatrick had appealed once more to the synod, and in October the presbytery again heard Kirkpatrick, cleared his reputation, and refused to honor Orr's request for a certificate of good standing.[38] As of April 1736 Orr, who was still owed the major portion of his salary, sued the congregation, and while the presbytery judged this action irregular and un-Christian, they nonetheless advised the people of Nottingham to voluntarily go before the magistrates and bind themselves for the amount owed. Orr's behavior continued to deteriorate until he received a presbyterial censure. That summer he fled presbyterial discipline and went to London where he was soon ordained as a deacon in the Church of England. The final note on this case recorded the settling of a new pastor at Nottingham in October 1736; the congregants were instructed to pay the portions due Orr, but there is no evidence that payment was ever made.[39]

This detailed account of a two-year pastor-congregation dispute illustrates the extent to which congregants would press their case when they felt themselves ill-used. It also reveals the ultimate power that the laity enjoyed. Despite the acquittal of Orr at the presbyterial and synodical levels and despite extended efforts to work through an accommodation, the disgruntled lay members managed to rid themselves of Orr. They simply stopped his salary so that he asked for the dissolution of the pastoral relation. Kirkpatrick appealed his case and pressed his charges until he cast enough doubt to prevent Orr from receiving a certificate. Life was made so unpleasant for Orr that, right after his certificate was refused, he left the country. All this he suffered for, apparently, preaching like an Anglican instead of like a Calvinist and for refusing to heed the wishes of his most influential, or at least most determined, congregants.

Unlike the prosecution against William Orr, the trial of Samuel Hemphill met with little resistance from the clerical leadership. Hemphill was also from Ireland, in this case a minister who had been ordained to go to America. In 1734 he arrived with a certificate from Ireland and was, after serious discussion with the First Presbyterian Church of Philadelphia over their parish commitments, permitted to accept an offer from Philadelphia to assist Andrews. An effective, gifted speaker, Hemphill soon filled First Presbyterian with hearers not usually in attendance at the church. Even Benjamin Franklin reported enjoying Hemphill's sermons, finding him a welcome change from the stodginess of Andrews. While Andrews may have been envious, other Presbyterians grew concerned as well, for Hemphill was preaching the moderate theology of moral virtue and human holiness. Within a few months he was brought before the Synod of Philadelphia and tried for his heretical views.

The trial created such a furor that repercussions were felt in Ireland. The Presbytery of Strabane in 1735 brought charges against Patrick Vance at the Sub-Synod of Derry for writing letters to America charging Hemphill with preaching unscriptural doctrines while in Ireland. Vance's defense was the truth of the accusations, and he proved, by means of several witnesses, that Samuel Hemphill had explicated Matthew 7:12 as follows: "who observes that golden rule, to do as they wou'd be done by, shall not miss their reward in time, and obtain eternall happiness hereafter." Hemphill supposedly said further that while a heathen's conversion to Christianity could be dated as a recognizable event, those who "led holy lives from their Infancy could not so exactly tell either the time when, or the manner how they were converted."[40] Vance was vindicated by the Sub-Synod of Derry and, later, by the General Synod of Ulster. At this latter hearing, Vance further added that he had written his letter

> at the desire of an inhabitant of America then with him, with a view to prevent the mischief which might accrue to that Infant Church by Mr. Hemphill's preaching such Doctrines there as Mr. Vance says he had been informed sd Hemphill had preached in Ireland. . . .[41]

Of course the real issue in Ireland was the reputation of the Presbytery of Strabane, considered by many a rather progressive presbytery, since that body had originally ordained Hemphill. By the time of the hearing by the General Synod, the Strabane ministers were reduced to complaining that Vance had not raised his charges before Hemphill's ordination was performed, an especially dangerous omission as his ordination was to a missionary charge and not to a congregation under the supervision of the General Synod.

The trial of Hemphill also served to unite the Presbyterian community against the common fear of outside heresy and the prevalence of free thinkers.

> Never were Ministers of the Gospel more loudly called upon, to appear in the publick Defense of our Holy Religion than in this dark Day of Apostasy and Infidelity, when the ancient Doctrines of Christianity are openly insulted and blasphemed, and everything that is sacred and venerable exposed to Scorn and Contempt.

So Ebenezer Pemberton of the New York Presbytery began his sermon before the commission ready to examine Hemphill. Pemberton went on to discuss the Christian faith as expressed by Calvin in opposition to non-Christians. He spoke on the general importance of sincere faith, purity of conversion as reflected in behavior, and attendance at worship, aspects common to all church members. The enemy comprised those who "can see no excellency in the Ordinances of Gods House, and esteem all Pretence of Communion with God in them, to be no better [than] the Enthusiasm and the Effects of heated Imagination."[42] The perceived danger must have been great to bring divines like Pemberton and Dickinson together with George Gillespie, who published his *Treatise Against the Deists, or Free Thinkers* at the same time.[43]

The April synodical commission had no problem finding Hemphill's theology heretical. The commission membership comprised ministers of all different backgrounds, for example, John Thomson, George Gillespie, John Pierson of Woodbridge, and William Tennent, Sr. The charges brought against him read very like the creed of an enlightened Arminian: Hemphill preached that there was no necessity of conversion if one had been raised Christian; he spoke against the doctrine of Christ's merit and satisfaction; he opened the church doors wide enough to admit all honest heathen; he described saving faith as assent to the gospel on rational grounds. Hemphill himself was quoted thus by a friend and advocate. "The doctrines absolutely necessary to be believed are so very plain and nigh unto us, that they are, as to their ultimate and most essential Parts, implanted in our very Nature and Reason."[44] No wonder that the Presbyterians were horrified; no wonder that Franklin so defended him. Of course, it is likely that Franklin's outspoken defense of Hemphill did even more harm, since the aspects that Franklin praised were the very points about which trouble had started.

> For the strain of Christian Charity that run thro' the whole of them, and their constantly urging the Necessity of a holy Live and Conversation in order to our final Acceptance with God [Hemphill's] sermons' were approved by People of all Persuasions. . . .[45]

Was this not a major complaint waged by orthodox Calvinists against the rational parties in Scotland and Ireland?

Perhaps the most likely cause for the unanimity, speed, and dispatch with which this case was handled also lay, understated, in Franklin's words. Phrases describing Hemphill's followers as "People of all Persuasions" and "Men of Sense" hinted at a universality that simply did not agree with ideas of election. The Presbyterians, as only one among many religious denominations in the middle colonies, were hard pressed to set themselves apart from other churches. In some areas the Presbyterians even had to share a building with the Anglicans.[46] When Hemphill began preaching in Philadelphia, he appealed to listeners outside the Presbyterian pale, and the Presbyterian congregants began to object. Rather than being seen as a benefit, a good evangelical effort, the presence of outsiders threatened the integrity of the community. "Most of the best of the people were soon so dissatisfied that they would not come to meeting. Freethinkers, deists, and nothings, getting a scout of him, flocked to hear."[47] Soon the rational theology would infiltrate solid Calvinism, and the purity of the church would cease. One of Franklin's most vehement points of argument invoked freedom of religion, which apparently meant that Hemphill should have been free to do whatever he thought best and not be cast from the communion. Dickinson neatly reversed this very point, explaining that freedom of religion gave the Presbyterians the freedom to refuse communion with anyone who they believed espoused questionable doctrine.

> There now appears the greatest danger that *Liberty* will be abated to *Licentiousness,* and that to escape *Imposition*, we shall open a Door to *Infidelity*, and instead of Charity and mutual Forbearance, we shall make *Shipwrecks of the Faith* as well as Peace of our Churches. . . .[48]

Samuel Hemphill rejected synodical authority; he was deposed. Later it was demonstrated that he had plagiarized the writings of others for his sermons, and he quickly lost his following among Enlightenment men and passed into obscurity.[49] Yet his brief presence in Philadelphia and his trial by the commission changed the developmental course of the Presbyterian church. The 1735 Synod meeting in the wake of his trial passed several resolutions to guard the church against the migration and employment of unworthy, unorthodox Irish students. Rigorous systems of trials and examinations were established in the hope of sifting out the heterodox.[50] In 1736 the synod, in response to doubts expressed by the people of Paxton and Derry, unanimously reaffirmed the Westminster Confession and their communal and individual subscription to it. The outside threat served to unite the ministers, at least for a while, and their congregants in protection of their self-conscious, elect community.

The pure, unadulterated community remained a central goal for the Presbyterians, not only the need to keep foreign elements out, but the hope that the community would continuously strive to purge itself from sin. To this end the congregations continued to hold large communion rituals. Several historians have noted the frequency of these rituals in the colonies, the several-day duration of the events, and the gathering of all neighboring congregations for a single celebration. Moreover, congregations still used testimonials and tokens as means of ensuring that the participants included only the worthy.[51]

In 1739 a Boston printer published a collection of *Sermons on Sacramental Occasions* by Gilbert Tennent, William Tennent, and Samuel Blair.[52] The construction of these sermons indicates a clear continuation of the communion ritual from Ireland. Each of the ten sermons is identified by date and use in the service, and these headings describe a service of at least three days, with at least three ministers present. The content and structure of the sermons themselves so closely resemble those of similar works by Irish ministers that there can be no doubt that this set of colonial services followed the Irish model.

The first two sermons were delivered consecutively before the communion. The first was an invitation to participate, emphasizing first the benefits of the sacrament and then the sin of staying away. Refusal to participate showed "rebellion" and "ingratitude" as well as "Unkindness to God and barbarous Cruelty to our Own Souls." Remember that it was necessary for the community that all but the openly sinful take part.

> Therefore let every honest experienc'd Persone, come to the Table of the Lord; in Obedience to his Command, that they may profess his Name before Man and Angels, and renew their Covenant with him here, in order to enjoy him hereafter.[53]

The second preparatory sermon was much more like the Irish sermons. This exhortation took its hearers through the conversion experience.

Gilbert Tennent began with his listeners' sense of sinfulness, self-knowledge of their inability to change, and the discovery of Christ. The truly converted then labored after Christ, closed with him, and worked to preserve their enjoyment of grace. Tennent ended his sermon with the warning

> that such as have not experienced the foresaid Characters of a true Claim to the Riches of Christ, wou'd not venture to come to the Lord's holy Table, in their present Condition, least they eat and drink judgment to themselves.[54]

Although this appears to contradict the first sermon, in fact the two together covered alternate aspects of the same process. All good Christians had to join in the communion of the congregation; any who were not holy were debarred as bringing defilement.

On this same occasion Samuel Blair preached the sermon during the distribution of the sacrament, that is, one of the Sunday sermons. His emphasis was the scheme of Christ's atonement, the great love that brought salvation to sinners. The sermon was more like a meditation upon Christian theology than an exhortation to its hearers. The communion day sermons were often of that quality—meant to give participants further, prayerful stimulation toward a joining with Christ, rather than to push forward a course of action.[55] Finally, there were the Monday-after thanksgiving sermons, in this case well represented by William Tennent's piece. Gone were all references to original sin, sense of affliction, and need to change one's ways. Gone also were abstract paragraphs on the images, metaphors, and meaning of salvation. Instead, Tennent reminded the participants that it was "the Duty of all those that have received the Lord Jesus to walk in him." People should strive to follow the commandments deliberately and constantly, and be warned against backsliding. Now that the conversion was complete, there was little left to do except keep the people steadfast in their purpose.[56]

One small incident in the Presbytery of Donegal illustrated well the centrality of the communion service and the care and peace that the laity expected to surround it. In May 1735 members of Thomas Creaghead's congregation complained that he had refused to allow his wife to partake of the Lord's Supper. The sessions had indeed met that Saturday morning and dealt with scandal in the congregation, but the pastor's wife was not presented to the session. Creaghead claimed that her sin had been committed Saturday night, that it was between himself and her only, and that knowing she had sinned, he, in conscience, refused to communicate with her. No one ever said what happened that Saturday night, but in the several examinations carried on by the presbytery over the next fifteen months, it became evident that as a second wife she resented an adult stepson who was invited with his family to spend time at Creaghead's homestead by Creaghead without her consent or knowledge. Various individuals were vindicated and censured, and the family was eventually reconciled.

What is important, however, is that Creaghead, because of his angry indiscretion, was ousted from his congregation. The elders had been so offended, the congregants were so convinced that the community was permanently scarred, that Creaghead was removed. After undergoing public repentance he was restored to the ministry, but the presbytery never entertained any hope of returning him to that people. Even when he was sent to them simply as a supply preacher, they refused to hear him. Six months after Creaghead's formal removal, the congregation was instructed to publicly acknowledge their fault in refusing to hear him. They did so gladly and asked for another supply; the presbytery did not send Thomas Creaghead. The affront to God and especially to the community had been overwhelming.[57]

Congregants maintained their own standards for clerical and lay behavior. These standards were more conservative, more flexible, and more difficult than those that the ministers had for themselves. The laity had certain ideas about orthodoxy which had to be honored. In one of his pieces, Franklin mocked Cross's indecision at Hemphill's trial:

> Tis strange a Gentleman of his acute Penetration cou'd not till after much Consideration discover Heresy in a paragraph that shock'd an illiterate Evidence at first Hearing and oblig'd him to run out of the Church in the middle of it.[58]

This made perfect sense and, incidentally, repeated previous patterns of lay response to unfamiliar preaching. Not knowledgeable on the subtleties of theology, the unlearned listeners latched onto phrases or catchwords that symbolized a position. The educated clergy could be expected to consider carefully the complete meaning as well as the rhetoric, whereas time and again the laity often responded virulently and negatively to preaching that was judged as poorly expressed but not unsound.

Tension between the clergy and laity would never cease. Whenever ministers felt free to pursue their own efforts, they were able to effect understanding, compromise, and unity. The subscription controversy was a prime example of this capability, since the laity remained uninvolved and two factions with very different views of polity were able to reach an accommodation. When the laity did become involved in church politics, it usually concerned an attempt to purify the church community by ostracizing and removing undesirable elements. The laity proved themselves remarkably able, through patience, fortitude, and a cunning knowledge of

Presbyterian procedure, to achieve their own goals, even if the clerical authority disagreed. Fortunately for the growth of the early Presbyterian church, this disagreement happened rarely. In fact, in the mid-1730s the laity and the clergy bound themselves together into an efficient, expeditious unit that strove to keep the unclean elements out of the community.

If any single factor could have been said to characterize this early Presbyterian community, it was the search for its own identity. All through the early years the ministers worked through problems one at a time, unsure of procedures and authority. Ministers from very different Calvinist traditions were tentatively approached and invited to join. With new members there came regularly new means of defining and structuring the church. When finally numerous enough to feel themselves an association, the ministers uncovered a pressing need to define, once and for all, what kind of church this was. Representatives from the various traditions argued and published, striving to make their own traditions the colony-wide traditions. The struggle began in 1721, and by 1738 the clergy and laity were united in their opinion of what the Presbyterian church was not. The problem of what the church was, what the traditions were to become, had yet to be resolved. This process of resolution occupied the colonial Presbyterian church for the next two decades.

6

Piety with Reason

The colonial Presbyterian church of 1738 was far from stable. The 1730s experienced a high level of Irish immigration causing an amazing growth spurt in church membership, a membership already too large and dispersed for the relatively few ordained ministers. In the eight years following the adoption of subscription three presbyteries, Donegal, Lewes, and East Jersey, and eighteen ministers were added to the rolls. The community had struggled through a major conflict relatively unscathed, and the members had finally bonded together in a show of unity in 1736. Still all was not well. Other problems were running just beneath the surface: personal disagreements among ministers, their beliefs, and their qualifications. Thus far these problems had been overwhelmed by the external threats to Presbyterian integrity. Soon, however, the size and progress of the community itself would remove these threats, and, as it did in Ireland in 1720, the Presbyterian church would turn inward, with vying factions insisting upon their own viewpoints.

The clergy essentially divided into four groups along a single continuum. At both ends were found extremists unwilling to compromise; in the middle were two reconciling, moderate parties. These two middle groups initially worked together to negotiate peace between the extremists. When a reconciliation proved impossible and the synod divided, the moderates themselves split, with some joining each new synod. All groups believed themselves to be continuing the godly, Presbyterian tradition, and both extremist groups fought to cast the impure elements out from the community. Into these battles the laity plunged, with some of the clergy urging full lay participation. During the next few years, despite intraclerical power struggles, the challenge of the renewed commitment to self-purification would move the laity to high levels of religious experience and engender a return to the days of the communion at Shotts and the Six-Mile-

Water Revival. The middle-colony revival would connect with similar forces in New England to constitute the movement now called the Great Awakening.

The Great Awakening within the Presbyterian community in the middle colonies has generally been perceived by historians as a brief, five-year crisis. Certainly from an institutional perspective, the two years of factional struggles that accompanied George Whitefield's tour followed by the split in the Synod of Philadelphia provide an easy target of scholarly concentration. The focus, however, should not be upon these critical points of institutional history but upon the phenomena that surrounded those events. Contemporary accounts indicate that the old Scots-Irish revivalist tradition began to surface several years before Whitefield arrived. New jealousies also appeared among Presbyterian clergymen, as well as hints of new alliances. No one can deny both the intrinsic and symbolic importance of Whitefield's early tours and the institutional crisis that followed. Nevertheless, broader exploration of religious patterns during the preceding decade will explain the synod's incredibly quick progress from tranquility to schism, as well as demonstrate the significance of this schism for contemporary believers.

During the 1730s, the debate over subscription had been closed off in favor of anxieties over the education and qualification of ministerial candidates, different perspectives that grew directly out of different beliefs about the importance and character of religious experience. One group favored the same practices, as did the General Synod of Ulster and the evangelical party (not the extremists) in the Church of Scotland. They called for the academic preparation of candidates, proved by credentials, and for a somewhat intellectual religiosity. As preachers they encouraged listeners to study the Bible and the Westminster Confession and Catechisms, to educate their children, and to live holy, moral lives in the sight of God. They did believe that individuals must experience conversion, but their Calvinist emphasis upon election discouraged them from seeking it. God in his own time would save his elected ones. The opposing clerics, called New Lights by their opponents, emphasized the importance of true piety for ministerial candidates. This was part of a focus upon emotional religion, with preachers deliberately provoking their hearers to travel through the throes of conversion. Also orthodox Calvinists, they nonetheless believed that it was an elected person's duty to strive toward conversion, much as the other side encouraged a holy life.

The two parties competed for the support of their clerical colleagues and the laity, although there was never any real doubt about where lay sympathy was. From the beginning the congregants rallied around the New

Light ministers. As George Whitefield toured the colonies in 1740, he heightened lay involvement so that, by 1744, when the synod finally divided, the New Lights had enough support to establish their own organization. Throughout these crisis years, the laity continued to promote their own active participation in religion, frequently and effectively challenging clerical authority when their own needs were not served. The history of the synodical schism over Presbyterian revivalism demonstrates that while the clergy were constantly adjusting their theology and developing their methods to fit the demands of a new environment, the laity had scarcely changed from their original commitment to emotional piety and revival.

By 1738 the ministers had realigned themselves along these new issues. Some continuity from positions on subscription might have been observed, but that was mostly coincidental. Both pro- and antisubscription parties had comprised several smaller networks, each sharing a cultural background and regional residence. These smaller groups remained largely intact in 1738, so that it was these groups, and not the individuals, that rearranged themselves around the new issues. Moreover, while the subscription controversy had split the community according to cultural identification, the conflicts of the late 1730s and 1740s cut across these cultural barriers. The leaders of the Old Light contingent included John Thomson from northern Ireland, while the New Lights were led by the almost dynastic clerical family from Antrim, the Tennents.

The Tennents were probably the single most important clerical force in the progress of the Great Awakening. William Tennent senior, the patriarch, was born in Ireland, educated at Edinburgh, and, in 1704, ordained in the Anglican Church of Ireland by the bishop of Down. From the beginning he seemed uncomfortable with Anglicanism, for he never served in a parish church, but only as a private chaplain to an Irish noble. Moreover, he had married the daughter of Gilbert Kennedy, a well-known minister in the Ulster Presbyterian network. He migrated to America in 1718 and immediately sought out the Presbyterian synod. They demanded his reasons for leaving the Anglican communion. Tennent listed several objections, primarily concerned with the antiscriptural government and discipline of bishops and their Arminian theology; his statement satisfied the synod, and he was accepted into membership. Initially settled in East Chester, New York, he moved to Bedford in 1720 and finally was called and established at Neshaminy in Bucks County, Pennsylvania.

Within the first year of his tenure at Neshaminy, in 1726, Tennent opened a seminary for the education of ministerial candidates. Known as the "Log College" by its detractors, this seminary provided the only education available to ministerial candidates in the middle colonies. Tennent was proficient in the classical languages and well read in divinity, and he quickly began to prepare a small coterie of devoted students for the ministry. These graduates of the Log College were hearty supporters of the Great Awakening; they, along with Jonathan Dickinson, were the movement's most able preachers and polemicists. The most important of these Log College men was Tennent's eldest son, Gilbert.[1]

Though he was actually ordained before the school's founding, Gilbert Tennent was nevertheless educated by his father and therefore has been identified as a "Log College man." This education was good enough to allow him to study at Yale, where he was awarded a master's degree in 1725. Licensed in 1725, Gilbert was ordained to the pastorate of New Brunswick in 1726. From the early 1730s he was an outspoken advocate of the centrality of evangelical piety rather than intellectual learning. He was an early, ardent supporter of Whitefield and in 1744 became the pastor of the congregation in Philadelphia that had grown out of Whitefield's audience. A leading, impulsive schismatic in 1741, Gilbert Tennent was also a leading proponent of reconciliation as early as 1748. He was effective in whatever cause he chose to espouse, and everywhere he traveled he enjoyed a large popular following. Gilbert did not rest upon his pastoral successes in New Jersey and Pennsylvania but embraced the role of itineracy. He was, next to Whitefield, the most successful preacher that toured the northern colonies.[2]

William and Gilbert Tennent were assisted by William's other three sons, William junior, John, and Charles, and by several Log College graduates, notably John and Samuel Blair and Samuel Finley; but the father and his eldest son outshone the others. Yet the two had very different personalities, resulting in decidedly different styles of leadership. William Tennent seems to have been a serious, competent scholar devoted to the education of ministers. He appeared to accept his own scholarly limitations in that he did not object to the synod's decision to put candidates who had no baccalaureate (presumably his students) through an examination process in order to guarantee their level of academic competence. Whitefield pictured him as "one of the ancient patriarchs. His wife seemed to me like Elizabeth, and he like Zacharias; both, as far as I can find walk in all the ordinances and commandments of the Lord blameless." Gilbert, on the other hand, was "a son of thunder who does not fear the faces of men."[3] Lacking the humble, grey-haired dignity attributed to his father, Gilbert was a presence to be dealt with: "Taller than the common Size, and in every way Proportionable," "a Man of *great Fortitude*, a *Lover of God, ardently Jealous for his Glory*, and *anxious for the Salvation of Sinners*."[4] When the qualifications of his father's students were questioned by the synod, Gilbert Tennent launched an attack.

During the Synod of 1738 the Presbytery of Lewes presented an overture on the private examination of candidates to the ministry. Merely commenting on the lack of formal education available to young men in the middle colonies, the new resolution required that any student who had not undertaken the usual curriculum in a New England or European college be examined by a learned committee, specifically appointed by the synod, before he was accepted on trials. Outside of its context, a close reading of this long, flowery overture reveals no factional spirit, only a concern that

> ys will fill our Youth with a laudable Emulation, prevent Errors young Men may imbibe by Reading without Direction, or things of little value, will banish Ignorance, fill our Infant Church with Men eminent for Parts and Learning, and advance the glory of God. . . .

Yet not once was Tennent's seminary mentioned in the minutes, nor was he appointed to one of the two committees of examiners. Of all the Log College men, only Gilbert Tennent, also sporting a Yale degree, was assigned.[5]

At this same synod, Gilbert Tennent and his sympathetic colleagues managed to organize themselves into a working organization. The Presbytery of Long Island had reported that their membership was so small that a quorum could not meet regularly to conduct business; thus they were joined to the Presbytery of East Jersey, which was renamed the Presbytery of New York. This allowed Gilbert Tennent and several others to leave East Jersey without crippling its organizational capability. This party, with a few from the Presbytery of Philadelphia, asked to be formed into a new presbytery, which supplication was granted, and the new erection named the Presbytery of New Brunswick. Despite other presbyteries' concerns about numbers, the New Brunswick ministers were happy with their membership of five. They were assigned a huge territory in northeastern Pennsylvania and eastern New Jersey.[6]

Excepting William senior and Charles Tennent, who stayed in the Philadelphia and New Castle Presbyteries, respectively, the Log College men were now united under a single organizational structure necessary to train and initiate new ministers. As George Whitefield noted during his visit to Neshaminy,

> It happens very providentially, the Mr. Tennent and his brethren are appointed to be a Presbytery by the Synod, so that they intend breeding up gracious youths, and sending them out into our Lord's vineyard.[7]

Whether or not this happened providentially, the intention of New Brunswick was clear. They had left behind in the Presbytery of Philadelphia the most avid opponents of the Neshaminy school, carrying over only a lot of personal animosity and bitterness. The New Brunswick members had determined to license their own candidates, and they began with John Rowland.

At its very first meeting in August 1738, the presbytery proceeded with Rowland's trials. The ministers considered the act of synod requiring a synodical examination before trials began and concluded unanimously "that they were not in point of Conscience restrained by sd Act from using the Liberty and Power which Presb^ys have all along hitherto enjoyed," and that it was their duty to take on Rowland.[8] The presbytery pulled this justification out of the Westminster description of church polity, which allowed presbyteries full control over ordinations. On the other hand, the New Brunswick Presbytery was definitely breaking with church tradition. Both the Scottish and Irish churches openly recognized the superior authority of the synod and/or assembly in all matters, including the licensure and ordination of candidates. The invocation of the Westminster Confession worked, essentially, to out-Presbyterian the Presbyterians. New Brunswick had contributed an innovative and more literal interpretation of polity that intellectually could not be answered.

Because John Rowland became the center of a qualifications dispute, it might initially appear desirable to evaluate Rowland's academic competence and compare his abilities with those of other ordinands. Yet such an assessment is not possible, for the moderates among the Presbyterians, the only ones who might be judged unbiased, never recorded an opinion of Rowland, either in official minutes or in print. A quick reading of Rowland's own published account of the revival in his congregation demonstrates a literary competence equal to that of the generality of publishing clergy. He successfully undertook the program of the Neshaminy school, and Gilbert Tennent did note that "he was competently qualified for the Ministerial Work in Respect of natural and acquir'd Endowments. . . ."[9] Of course, what else would Tennent say at the funeral of a Log College man? Be that as it may, debating Rowland's competence serves no immediate purpose and, in the context of the synodical dispute, was irrelevant. Rowland was a pawn, albeit a strong and willing pawn, in New Brunswick's planned confrontation with the synod. Any candidate from the Log College would have served the presbytery's purpose, and I believe the synod majority would have challenged any candidate. That Rowland was a particularly gifted preacher would only make matters worse.

The month after Rowland was licensed, the congregation of Hopewell and Maidenhead, having no pastor, requested that Rowland be permitted to supply. On the border between the New Brunswick and Philadelphia Presbyteries, this congregation fell under the formal jurisdiction of Philadelphia. The previous spring the Philadelphia Presbytery had sent John Guild, their probationer. The people claimed their right to hear a second candidate and asked for Rowland. The presbytery tried to obstruct this move by allowing the congregation to hear any *regular* candidate. Notwithstanding this restriction, the congregation invited Rowland, who agreed to preach. David Cowell of Philadelphia told Rowland that his attendance would cause dissension. Still Rowland went, and indeed catalyzed a congregational split. In October the Presbytery of Philadelphia allowed Maidenhead and Hopewell to be divided into two congregations. Hopewell called John Guild, while Maidenhead asked to be removed to the jurisdiction of New Brunswick. At its next meeting, in May, the synod reproved New Brunswick for admitting Rowland to trials without the proper examination; further, the synod judged that the congregation of Maidenhead had

> behaved with great Indecency toward their Presbytery by their unmannerly Reflections and unjust Aspersions both upon the Synod and the Presbry, and yt they have acted very disorderly in improving Mr. Rowland as a Preacher among them, when they were advised by the Presbry yt he was not to be esteemed and improved as an orderly Candidate of ye Ministry.[10]

During the same synod, an overture for the establishment of a "School or Seminary of Learning" was approved, and a committee that did not include any Tennent or any member of the New Brunswick Presbytery was appointed to this task. In August 1739 this committee met with correspondents from every presbytery and with the synod commission, an advisory body that acted for the synod during the year. No correspondent represented New Brunswick, nor did Gilbert Tennent, a synodical commissioner, deign to come. Instead, New Brunswick continued its own efforts, ordaining John Rowland to an evangelical ministry in November 1739. As if to make a halfway gesture of reconciliation, the Synod of 1740 "clarified" its decision concerning the acceptance of privately educated probationers without examination, explaining that it did not question the right of subordinate presbyteries to license and ordain ministers, but merely asserted the synod's right to evaluate candidates for membership into the synod. Thus Rowland was allowed to preach within the bounds of the Presbytery of New Brunswick while the synod could still save face.[11]

This tale of turmoil between the New Brunswick Presbytery and the synod should not be read as the oppression of a small, godly minority in the face of an unwavering, arbitrary institution.[12] True, the synod had backed the New Brunswick members into a corner, passing several resolutions in the wake of New Brunswick's opposition. It is also true that one or two questionable candidates from the Presbytery of Donegal had been ordained against the opinion of other ministers. An outstanding example is the case of Richard Sanckey, discovered to have plagiarized a discourse delivered during his trials. Even worse, Sanckey had researched works that contained "gross and evident Errors," and then he had forwarded his notes along to Henry Hunter, another probationer, for his trials. Donegal had only rebuked him and stopped his trials for a short time, a decision that was upheld by the synod despite a remonstrance presented by George Gillespie.[13] Still, the Log College ministers were not without defenses. After all, the new presbytery had been requested and organized before the other measures concerning ministerial qualifications had been proposed. The synodical examination overture may, in fact, have been a response to the new network.

The Log College men also had definite opinions on ministerial qualifications. Less worried about academic achievement, they wanted a candidate to provide evidence of the gracious state of his soul—evidence of his conversion. These were the terms upon which the New Brunswick Presbytery accepted probationers for trials, and since all had found Rowland to be a fine young man of exceptional piety, Rowland was considered fit to begin the process. Gilbert Tennent had for a long while espoused the cause of a gracious ministry; and, in the wake of the activities of the Donegal and Philadelphia Presbyteries, he took his cause to the people. On the eighth of March 1740, Gilbert Tennent entered the pulpit of the Nottingham meetinghouse and preached on *The Danger of An Unconverted Ministry*.[14]

No one reading this sermon can believe that Tennent's party was at a disadvantage. Fast and furious, the rhetoric hardly left the topic of the wickedness and evildoing of the unconverted minister. The Log College men had wearied of avoiding illegal actions and began to proclaim their own style of ministry. It was important, too, that Tennent chose Nottingham as the pulpit for his exhortation. The laity here had already demonstrated a marked independence from the Presbytery of Donegal in their prosecution of William Orr. More recently, the congregation had been vacant for several months, and, after patiently awaiting presbyterial supplies that never came, the congregation had invited and heard Samuel Blair. Donegal had, of course, objected, but again the presbytery had little power over a congregation that it could not supply. This audience was primed to hear the definitive defamation of its own clerical authorities.

After labeling the unconverted ministers as "Caterpillars" who "labour to devour every green Thing," Tennent devoted several minutes to developing the metaphor of Pharisee, squarely placing the unconverted ministers within it. The Pharisees were "proud & conceity," "crafty as foxes" with the "Cruelty of Wolves"; they "had their Eye, with Judas, fixed upon the Bag." Tennent described them as fierce bigots enamored of the small, unimportant rules of religion. "The Pharisees were fired with a Party Zeal. . . ." Indeed, they had no redeeming qualities whatsoever, since only by conversion could an individual do good. By definition, therefore, all unconverted men were evil.[15]

After devoting a very few minutes to justifying scripturally the necessity of rebirth through the story of Nicodemus, Tennent returned to the flaws of the unconverted preacher—now the weakness and ineffectiveness of his sermons. Discourses were "cold and sapless, and as it were freeze between their lips." They often provided security to the wicked, afraid to challenge any evildoing, and—the common complaint levied by evangelical Calvinists against their opponents—they confused legal obedience with gospel obedience. Thus they implied that individuals by obedience could achieve salvation. "They keep Driving, Driving, to Duty, Duty, under this Notion that it will recommend natural Men to the favour of GOD, or entitle them to the promises of grace and Salvation. . . ." Finally Tennent cited the congregations of unconverted ministers, which he found singularly irreligious and dead, and in need of the labor of a converted minister. And if there was any doubt that Tennent was espousing a personal cause as well, it was effectively removed by his conclusion to the exposition portion:

> The most likely Method to stock the Church with a faithful Ministry, in the present Situation of Things, the public Academies being so much corrupted and abused generally, is, To encourage private Schools, or Seminaries of Learning, which are under the Care of skillful and experienced Christians; in which those only should be admitted, who upon strict Examination, have in the Judgment of a reasonable Charity, the plain Evidence of experimental Religion.[16]

Tennent devoted equal time to the improvement portion of his sermon, calling people sad, lost fools for staying with an unconverted minister, however well behaved, and encouraging people to leave their ungodly ministers. He even allowed people to leave good ministers of lesser abilities for those with greater gifts.[17] In other words, people were more or less licensed to hear whomever they pleased, wherever they pleased, provided, of course, that the sought-after preacher was godly. To grant the laity this much choice in the selection of a pastor not only threatened the livelihoods of many clerics; such allowance also undermined the sanctity of the pastoral relation as laid out in the Westminster Directory.

The perception of a minister as called and prepared by Christ was reinforced by the symbolic installation of a pastor to his charge. Some men were more talented than others, but the Calvinist's assurance had rung clear that if providence had placed a minister in a particular congregation, that was the foreordained field that would profit most by his labors: "if he be gifted and sent of God to you in particular, and by the Providence of God settled among you . . . then you cannot expect the same Blessing in hearing others as in hearing him."[18] When a disgruntled segment of the Neshaminy congregation tried to rid themselves of William Tennent, Sr., they used as justification the fact that the presbytery had never formally installed him. The synod, however, despite the majority's lack of sympathy with the Log College men, refused to allow this relation to be severed. A recognized pastoral relation could not be set aside for the mere lack of installation (although the synod did not approve this negligence); such was a sacred and serious relationship.[19] Jonathan Dickinson, no friend to the hierarchical judicatory system, preached vehemently against the laity's rejecting ministers who had been ordained, with the people's consent, to their care. Speaking in defense of John Pierson, whose congregation had been expressing dissatisfaction with his demeanor, Dickinson emphasized that such expression wrongly attributed to men the work of conversion. Conversion, he reaffirmed, was brought about by the Holy Spirit. "This mistake lies in giving the Honour to the Instrument, which belongs only to the principle Agent; and not ascribing to the Sovereignty of Gods free Grace, all the blessings that he is pleased to afford to the means of his Grace." He added that the rejected ministers supposedly "want right views, are not influenced with a Zeal for the Cause of Christ, or with a Love to the Souls of Men. They and their followers are *dead* and *lifeless*, of a Laodicean Indifferency," and Dickinson replied to such opinions with the challenge that he, for one, could not see into men's hearts, and he asked others "to refrain from doing the same."[20]

Dickinson's eloquence notwithstanding, the laity adhered to Gilbert Tennent's position. Within a few months his sermon underwent several reprintings in Philadelphia, including a translation into German.[21] The broad appeal of the sermon indicates a level of lay involvement not explained by a personal quarrel among clerics over the adequacy of a private seminary. Why were the laity so active? And even if Tennent's reaction could be attributed to the shabby treatment his father had received, the question remains of why the synod acted as it did. The Tennents were not opposed to an educated ministry, so they could not reasonably be accused of producing ignorant clergymen, as some English nonconformists did. Still, the synod implied that the Log College commitment to intellectual preparation was inadequate. Obviously something was frightening the ministers, perhaps the same something that so appealed to the laity.

All ministers were agreed that a pastor should evidence, as far as possible, an elect status. However, Presbyterians had generally rejected the possibility of determining without doubt a person's state of grace. Thus a man was evaluated for the pastorate in terms of his moral behavior and learning—the perceivable evidence of his call. In an early essay on the ministry and its relation to the laity, David Evans laid out the traditional qualities sought in a pastor, such as faithfulness, holiness, sound faith, and the ability to communicate. He listed learning and knowledge as of no small importance. "Whatever too many, by a fond Dream and Conceit, prattle against human Learning in a Minister of Christ, yet we are sure that Learning and Religion ever did, and do, fall or flourish together."[22]

Now the Tennents were suggesting that some investigation into a person's piety, his acquaintance with experimental religion, would test the truth of his conversion. Here was the old Protestant dispute over the nature of saving faith: was it characterized by right reason or an emotional conversion? Conversion was considered essential, but the Augustinian tradition had denied the possibility of distinguishing the converted from the unconverted. Thus most Calvinists, taking Augustine's model of conversion as their own, had turned to right reason and moral behavior as a general guide. Only a few dared propose the possibility of a visible church, but these few had included the large, strong, Puritan community in New England. And the Puritans' arguments apparently convinced Gilbert Tennent. The Log College men emphasized the importance of this conversion experience for service to the church. When Gilbert Tennent called the ministry unconverted, he was not merely impugning their fitness as ministers because they had not passed through a prescribed ritual. He was attacking their fitness as men. The unconverted minister was condemned for all eternity; no wonder that the accused were angry. Tennent quickly followed up his Nottingham appearance by presenting, with Samuel Blair, a statement lamenting the irresponsible careers and the lack of piety of many of the members. Unwilling to be drawn into combat at this time, the synod merely admonished all ministers and recommended that presbyteries more closely supervise their congregations.[23]

Had the real problems concerned only ministerial competence, perhaps the conflict would have eventually been resolved in synod, as it had been in 1729. However, the primary clerical anxiety involved not the ministers themselves, but the laity. The Log College men, with their emphasis upon conversion, had been stirring the congregations to active interest in religious happenings. Their position on ministerial qualifications was only one aspect of a religious system that excited the laity across the countryside, excluding nonsupporters from lay approval. From about 1730 onward, the Log College men had been nurturing revivals.

There was some evidence of revival as early as 1729. In 1726 Gilbert
Tennent, new pastor of the congregation in New Brunswick, met Dominic
Theodorus Jacobus Frelinghuysen. A Dutch Reformed minister, Frelinghuysen had arrived in the middle colonies in 1719 and within a few years
had become a central religious figure in the Dutch community. Impressed
by the effects achieved through Frelinghuysen's strict discipline and evangelical preaching style, Tennent felt the ineffectiveness of his own work in
New Brunswick.

> I began to be very distressed about my want of success; for I knew not for
> half a year or more after I came to New Brunswick, that any one was
> converted by my labors, although several persons were at times affected
> transiently.[24]

At this time, too, Tennent was severely ill for a prolonged period,
and, like a good Calvinist, he used that time to belabor his pastoral failure.
Upon recovery Tennent began his work with a new zeal, rallying his
congregants to a concern for their salvation. Whether the Dutch pastor,
through his advice and friendship, had actually taught Tennent techniques
or merely reinforced inclinations is unknown. However, it is evident from
statements like the one just cited that Tennent's theological framework had
already placed conversion at the center of religious experience.[25]

Gilbert Tennent actually credited the beginning of the revival to the
efforts of his brother John. Settled in the congregation of Freehold, New
Jersey, John Tennent in the two years he worked there (he died in 1732)
precipitated a revival that was continued by his brother and successor
William. His listeners believed John Tennent's assurance that "Regeneration is absolutely necessary in Order to obtain eternal Salvation," and they
accepted his warning that none could achieve regeneration through outward professions or behavior. Each person had to discover sin, suffer
anguish over the state of his or her soul, repent, ask what could be done
to seek salvation, and understand the renewal and assent to the change.
After this conversion, one's judgment was altered, values and goals were
changed; one's reasoning ability, will, and conscience were greatly strengthened. The congregation of Freehold struggled through its revival, and the
Great Awakening in the middle colonies had begun.[26]

All through the 1730s the Tennents nourished the revivals in their
congregations. Gilbert Tennent in 1735 published, in addition to the sermons of his brother John, at least five of his own revivalist sermons.[27] He
was a powerful, strident preacher, not afraid that his harshness might
alienate his people. A fellow minister described Tennent's effect on his
hearers thus: "Hell, from beneath, was laid open before him, and Destruc-

tion had no covering; *while the Heavens* above *gathered Blackness* and a Tempest of Wrath seemed ready to be hurled on the guilty Head."[28] He called for "religious violence," and employed images such as thunder and lightning, the leviathan, and chains of the devil. Essentially, these sermons were exceptional examples of the rhetorical structure and images so useful in provoking a congregation to an emotional awakening.

A Solemn Warning to the Secure World was perhaps the best known of the pieces published that year. The book opened with a thirteen-page preface encouraging individuals simply to wake up and consider the lost state of their souls.

> Will not the *Terrors* of an *eternal God*, and an *eternal Hell* make you *afraid*? . . . Are you *degenerated* into *Beasts*? Are you cover'd with the *Leviathan's Scales* that no *Arrow* from the *Bow* of *God* will pierce you? . . . But perhaps you mock at Fear and are not affrighted, though the *Heavens* look *black*, and *God's Lightnings* and *Thunders*, from *blazing, trembling Sinai, flash* and *groan* and *rore hideously!*[29]

Tennent then proclaimed the threat of judgment by law and reasserted the commonly known truth that all were condemned according to the law. It was only after several more pages of attention-getting rhetoric, including several paragraphs beginning "Awake Awake," that Tennent cited his biblical text and began the actual sermon.

Tennent used for his text three verses from the twenty-ninth chapter of Deuteronomy, the chapter in which Moses delivers the covenant to the people of Israel, the Old Testament prototype of the elect. Tennent appropriately chose verses 19, 20, and 21; these verses have explicit warnings for those people who turn away from God. The first verse describes the self-delusion and complacency of those persons who find themselves righteous even though they have rejected the covenant. The middle verse declares God's curse upon these people, while the last warns that the cursed sinners will be separated from the tribes of the chosen. Thus, Tennent provided himself with an excellent support for identifying "confident, secure" sinners and threatening them with the worst of punishments. The chapter continues for eight more verses, chronicling the full indignation of God—verses that refer to brimstone and Sodom and Gomorrah—and retelling the reasons for the curse. Note also that these three verses are divided from the discussion of the covenant and rewards promised to the faithful that comprise the first part of the chapter. The means of this division are five verses of identification of the sinners. Tennent deliberately selected three verses immersed in threats and curses for smug sinners, allowing no interference of his terrifying theme from the promises made to the elect.

The running head for this sermon was "The Presumer Detected," and over and over Tennent provided many different images and structures for understanding the secure sinner's nature and destiny. Putting aside the backsliding of the converted Christian, Tennent reaffirmed his emphasis upon true wickedness. Because he wanted to confront those people who believed that they were living righteous lives, Tennent devoted some time to the ways that people hid their own sins within the cares of the world. "But when married many chuse another Master, they serve the World with much Fidelity and Care, and under the Pretence of providing for their Families, loose the Opportunities of Mercy, and so starve and damn their Souls."[30] He then dissected vices in detail and discussed the way that these worked to deceive the guilty ones, even to the point of false convictions that they had experienced the entire conversion process.

Tennent then found himself enmeshed in the Calvinist paradox: one could not effect one's own conversion; therefore why should any effort be exerted? He responded with the federal theology that had served the New Englanders so well. He told his readers that they could still prepare for conversion; "we can seek a Change after some sort, and if we do not we perish deservedly. . . ."[31] Somewhat uneasy with this solution, Tennent soon found an excuse to attack the opposite heresy, Arminianism, as restrictive of God's arbitrary will. While he was uncertain, like most Calvinists, about the need to work toward salvation, he had no hesitation in rejecting humanity's ability to achieve salvation, rather than receiving grace through the Holy Spirit.

As early as this sermon Tennent preached against unconverted ministers, blaming them as the source of false security and unsound doctrine. At this stage the unconverted ministers were an identifiable group, easily distinguished from all other Presbyterians by their theology of free will, universal grace, and universal redemption. Tennent was speaking against the Arminian theology espoused by the Anglicans and, worse, the Deists. This enemy was still the outsider, corrupting members by luring them away from true Christianity with false hopes of an easier salvation. Not until his own institution rejected his style of ministry did Tennent turn on his fellow Presbyterians and recharacterize the converted ministry so as to exclude his synodical opponents.

A final point was the varying character of Tennent's (and the other revivalists') preaching. They rejected the standard Presbyterian method of general preaching, by which a minister would preach on consecutive verses from the Bible until a chapter was completed. This method the revivalists considered unsound, since each soul was at a different point along the conversion process, and no one sermon could be appropriate for everyone.

Thus, in one sermon Tennent could preach about God the mighty judge as he warned sinners "that as God knows your secret Impieties, so he will damn you for them, he will *tear you in Pieces*"; in another he described Christ the husband and asked, "Have you consented to the Terms of the Covenant of Marriage. . . . Have you had Communion with Christ by his Word and Spirit?"[32] He could beg the sinner to forsake evil, threaten the sinner with future torments, encourage the hopeful convert, or reaffirm the joys of the saved. At all these preaching functions, Tennent and his associates were masterful, provoking intense, emotional, sometimes physical responses that would frequently lead, as far as they could tell, to complete conversion. Touring the countryside with their collection of terrors and joys, these new itinerants attracted more and more listeners. And as the laity displayed a preference for the revivalists, other Presbyterians became antirevivalist—some from envy of the revivalists' success, some out of sincere conviction of the errors of revivalism, and some out of fear for their positions.

The problem quickly became one of intrusions. In the formal presbyterial system of government, a minister held absolute religious authority over the territory embraced by his congregational bounds, just as a presbytery controlled a fixed geographic area. Since a pastor was fully responsible for all people within these bounds, he might also have expected a free hand with them, with no other minister undoing his work or competing for the people's attention and money. When a minister entered another's congregation, he was, essentially, trespassing. The visiting preacher was expected to ask permission to enter and preach, and in granting that permission (and permission generally was given), the pastor retained ultimate control over his congregation. If no pastor had been installed in a congregation, visiting preachers were expected to go through the presbytery, both to avoid duplication of effort and conflicts in time and to maintain discipline.

In the colonies, this structure worked far less efficiently than in Britain, due to the vast amount of space enclosed in a single congregational unit. One congregational charge might have two meetinghouses twenty miles apart, with families scattered about a ten-mile radius of each building. Not only was the minister's work full of tiresome, endless journeys, but people were so distant that many did not attend services regularly. Going one step beyond to the presbytery one would find ten, twelve, even fifteen such congregations, many with two or three gathering places, in a single jurisdiction. With the amount of work involved and the scarcity of clergy in the colonies, most presbyteries could not serve even half the preaching places in their territories. Because of this alone, most vacant congregations would have welcomed any intruder, for the people, by erecting congregations by themselves, demonstrated that they enjoyed religious services and sought out the sacraments. On the other hand, intrusions might have been perceived as an indirect criticism of the presbytery's failure to meet its responsibilities. Moreover, intrusion into a pastor's congregation would have been doubly resented, for there were plenty of preaching stations vacant and in need of a licensed preacher.

As early as 1737 the ministers began to grow concerned over the possibility of intrusions. No official document recorded any complaints, and no pamphlets against such behavior had appeared, but the Synod of 1737 passed a series of resolutions aimed directly at controlling the movements of probationers and the activities of congregations. Although there must have been some activity to precipitate such anxiety, not even the preface of the overture revealed specific perpetrators or victims. Rather, basic orderliness in church affairs was invoked, after which five restrictive resolutions were passed as a general reinforcement of church government according to the directory. No probationer could preach in a vacant congregation without the permission of both his presbytery and the presbytery within whose bounds the congregation was located. A presbytery could not invite a probationer to preach without a recommendation from the probationer's presbytery, nor could a congregation invite a probationer or minister without the consent of its own presbytery. Finally, no minister could invite any probationer or minister to preach in a vacant pulpit without the concurrence of his colleagues in the presbytery. Summarized with a reminder that all probationers, ministers, and congregations were under the authority of higher judicatures, this overture provided the guidelines by which intrusions would be identified and censured.[33]

By 1738 itineracy was openly recognized as a problem. The synod approved an overture denying ministers freedom to preach in a congregation in any presbytery "after he is advised by any Ministr. of such Presbry yt he thinks his Preaching in yt Congregatn. will have a tendency to procure Divisions and Disorders," unless the minister had the permission of the judicature. Still the synod firmly assured ministers that this resolution was not intended to interfere with a minister's preaching at some place he "providentially" happened to be, unless that individual had been informed of the likelihood of discord. The following year the synod reinforced the act against itineracy with the added stipulation that anyone who suspected a disorderly and discordant intrusion should report the intrusion to the congregation's presbytery, and the intruder would be obliged to appear before them.[34]

Thus not only did John Rowland flout the regulations regarding the academic examination of candidates, he also intruded into a congregation after he had been warned of congregational discord that might (and did) ensue. The anger against Rowland was exacerbated by the outbreak of a revival at Maidenhead that lasted almost two years. Rowland's own report, combined with the fact that half the congregation did remove itself to the care of New Brunswick, indicated his abilities as a revivalist preacher. Gilbert Tennent praised just these qualities.

> And indeed his Talent of convincing the Secure was uncommon, he repre-sented the Dangers of their doleful Case in a strong Light, in a dreadful Dress, and on his Account he might be justly call'd a Boanerges, or a Son of Thunder.[35]

The New Brunswick Presbytery thought Philadelphia's protest ridiculous, for the congregation had been vacant, and the people had themselves invited Rowland. The Log College men simply did not agree that because one minister, from as much as a hundred miles away, rejected a preacher as a cause of discord, the individual in question could not preach there.

The appeal of thunderous preaching to the Presbyterian laity was incontestable. While the orders against intrusions may have seemed unrea-sonable to the Log College men, the amazing effects of these intrusions upon the laity went far to justify such restrictions. The New Brunswick members may never have intruded into a congregation that had a pastor, but this did not prevent such a congregation from being affected by the fervor surrounding it. Even though no one had ever preached in John Pierson's parish uninvited, news of the revivals had spread to Woodbridge, bringing dissatisfaction to the people there. Six months before Tennent had delivered his *Danger of an Unconverted Ministry*, Dickinson castigated the people of Woodbridge for "a factious setting up and preferring one faithful Minister of the Gospel above the other. . . ." He mocked the lay preference for "the most thundring and terrifying Methods of address" and ridiculed the revivalist penchant for "frequent and open Discourse about their Con-victions and other spiritual Experiences." Yet while Dickinson did attack the style of the revivalists, his primary invective was saved for the con-gregants themselves:

> This is to promote the best Welfare of their Neighbours. This is to bring them under the advantage of a more Powerful Ministry, whereby their eter-nal Interest may be best promoted. 'Tis to associate them with more serious Christians; and to bring them acquainted with more vital Piety. These and such like Panegyricks upon themselves and their Party, are too commonly heard from some among us, who seem to suppose they are in the Service of Christ, when breaking in upon the Peace of the Churches, and actually endeavouring to foment Discords, Divisions, and Confusions among them.[36]

Donegal Presbytery experienced more problems than most from the intrusions, thus producing some avid opponents of the revival. Nottingham was, after all, the site of Blair's censured intrusion and Tennent's vituperative response. In September 1740 the congregation at the Forks of Brandywine brought several charges against their pastor Samuel Black, whose trial was held in November. The charges included seeking after his stipend, complaints of fatigue during pastoral visiting, "superficial regard to promote personal or family piety," intemperance, lying, and sowing sedition. The accusations of immoral conduct, especially intemperance, to which Black himself confessed, seemed to be valid.

A slightly deeper examination, however, revealed that the true dissatisfaction lay in Black's antagonism toward the revival. The people had recently enjoyed the uninvited preaching of revivalist David Alexander. The congregation explicitly requested that Samuel Blair and Charles Tennent, both of New Castle Presbytery, be invited as correspondents to judge the charges. Donegal agreed to correspondents from the predominantly antirevivalist New Castle, but would not allow the congregation to have either Blair or Tennent. Further, the accusation of lying involved Black's initial condemnation of field preaching, followed by a more positive appraisal in view of the apparent moral reformation of the participants, followed by renewed distrust. So too, his sowing sedition actually involved censures levied against congregants who supported the intruders. The last two articles, in fact, were summarized as "his Seeming to oppose the work of God appearing in ye land." The presbytery's judgment exonerated Black, although he was warned to be careful of the manner in which he presented himself to his people. Donegal considered the differing opinions on the revival ample justification for Black's wavering position. The presbytery also condemned the manner in which the laity had brought the charges, discovering much malice in the laity's actions, both in their eagerness to present copious oral and written evidence and in their failure to bring charges of long-standing offenses until Black had actually refused to condone or encourage the revival.[37]

Francis Alison, during this same autumn, complained to Donegal that Alexander Creaghead had intruded into his New London congregation (New Castle Presbytery), where he caused division and hostility among pastor, elders, and people. Since John Thomson's congregation at Chestnut Hill in Donegal Presbytery had also suffered from Creaghead's and David Alexander's intrusions, Donegal proceeded almost immediately to Middle Octara, Creaghead's congregation, to investigate the charges. Creaghead declined the presbytery's authority, judging the entire group as his accusers and therefore incompetent to judge. Alexander had also slighted the presbytery, scrupling to attend judicature meetings because the members opposed the work of God and condemned crying out during sermons. Moreover, Alexander felt that the presbytery performed only "superficial" examinations of ministerial candidates.

The exciting aspect of this trial involved the actions of the laity; the trial was frequently interrupted by an uprising of the congregation. The minutes noted that Creaghead and Alexander "industriously detain'd yᵉ Pby from doing business in an orderly way by breaking in upon us and consuming our time by circumlocutions & harangues to amuse yᵉ populace." When the presbytery came to the meetinghouse they found Creaghead preaching upon Matthew 15:14 ("[T]hey be blind leaders of the blind. And if the blind lead the blind, both shall fall into the ditch.") and delivering a continuous invective against "Pharisee preachers & carnal Min.ʳˢ." After the presbytery had begun the trial, Creaghead directed his congregation into a tent next to the meetinghouse where Alexander and Samuel Finley read Creaghead's defense, a paper that reproached several Donegal members by name. The following day the presbytery tried to proceed, only to be forced to listen to Creaghead's defense again. While Creaghead did apologize for the congregational antics that regularly interfered with the presbytery's business, he laid the blame at the feet of the ministers who refused to answer the charges he had raised. Since all the articles of accusation against Creaghead were either proven or admitted, the presbytery suspended him, while he declined its jurisdiction.[38]

By the end of the 1730s the Scots-Irish clergy had divided themselves into competing factions. Although several ministers, notably George Gillespie, remained uncommitted on the subject of the revival, far too many, for the peace of the Presbyterian church, took a position. Of course, all the Scots-Irish embraced orthodox Calvinism with an emphasis upon original sin, the covenant of grace, and election. All through his *Solemn Warning*, Gilbert Tennent discounted both Arminian and antinomian tendencies in favor of a staunch federal theology. Just one year before, David Evans, a determined opponent of the revival, had preached a sermon entitled *Law and Gospel* in which he demonstrated humanity's utter inability to achieve perfect obedience and the resulting dependence upon Christ's satisfaction and the covenant of grace. Preached in the point-by-point, textually supported method of the plain style, its purpose appeared to be to teach rather than to provoke.[39] Whether the preacher used inflammatory rhetoric or a twenty-item applications section, the theology was the same brand of Calvinism. Yet by declaring for or against the revival, a minister was owning an entirely different philosophy of religiosity.

The differences lay in the manifestations of their beliefs in religious rituals and the church's institutional structure. Those opposed to revival, the Old Lights, favored a highly skilled, academically trained ministry. Endorsing the formal, hierarchical structure of Presbyterian government, the Old Lights felt the future of the church was safest in the hands of the recognized authorities, in this case the clergy, with a few representative elders. Even the elders were granted little power in the actual affairs of the colonial church; their opinions were certainly never noted or acknowledged officially. This emphasis upon clerical authority naturally placed the pastor at the center of congregational life, with the laity as passive recipients of wisdom and guidance. This paralleled the theological concept of salvation as a free gift from God to the passive elect. Such religiosity was close to the philosophical position of the Church of Scotland. The colonial ministers' theology definitely did not follow the moral philosophy lines of Scotland's moderate party, but their views on ritual and their preference for a passive, docile laity were similar.

The prorevival ministers, or New Lights, called for an active laity with an entirely different focus for their religiosity. While none denied the necessity of an educated ministry, the New Lights emphasized personal piety as the central consideration. As it was with the clergy, so with the laity: they were not to sit and be taught but to participate actively in their own conversion. Through the explanatory framework of covenant theology and revival as a tool of providence, the New Lights emphasized the need for conversion as a primary tenet derived from Calvinism. In encouraging the laity to seek conversion, the revival leaders resembled the Seceders of Scotland. In fact, Gilbert Tennent, through his early career, corresponded with Ralph Erskine. And Alexander Creaghead, one of the problem ministers in Donegal, sympathized with the Secession Church, demanding that various persons and groups subscribe the Solemn League and Covenant. (No one would agree to this subscription at all, much less as a term of communion, and Creaghead left the synod.) Obviously the revivalists were not ideological colleagues of the Seceders, for the sins of Scotland meant little in the colonial environment. Nonetheless, in their establishment of the presbytery as the primary organizational unit, their emphasis upon emotional piety, and their encouragement of lay participation, the methods and rituals of the Seceders and the New Light ministers were very similar.

When the revivals first began, the laity had little opportunity to express a preference. The need for Presbyterian clergy was so great and the numbers so few, that most congregations were pleased enough to get anyone. The restriction of preaching to an educated ministry and the centrality of the sacraments to religious experience rendered the presence of an ordained minister important, and during a scarcity any minister was adequate. As long as he did not defile the sacraments by his own poor behavior or heterodoxy, he was acceptable. Later, however, when the journeys of itinerants provided the laity with a choice, the people had no problem choosing revivalism. As the itinerants toured the countryside, their very popularity threatened the status of those not involved. This was not unlike the situation in Ulster during the Restoration period, when the laity rallied in support of the Scottish itinerants, criticizing their own quiet and non-disruptive pastors. The occasional instances of the laity's effectively challenging the clerical authority during the 1730s demonstrated that with a concentrated effort they could alter the course of the Presbyterian church. At the end of the decade there lacked only a force to mobilize the lay support for the revival. In November 1739 that force came in the person of George Whitefield.

The escalation of the ministerial competition from haphazard outbreaks of reactions against the Log College men to wholesale efforts to repress their work was a direct reflection of the vast success that Whitefield experienced in the middle colonies. His reputation as an efficacious preacher and kindler of souls had preceded him, so that while Whitefield had planned to go directly to Georgia upon his arrival at Philadelphia, the Tennents encouraged him to spend some time preaching in their region. Whitefield did not need much encouragement; he generally lost no opportunity to exhort people, no matter who or where they were. "Went at the invitation of its father to the funeral of a Quaker's child," wrote Whitefield, "and thought it my duty, as there was a great concourse of people at the burying place, and none of the Quakers spoke, to give a word of exhortation."[40]

Even if Whitefield did exaggerate, his journals still indicate that wherever he went, he met a great concourse of people. Six thousand people in Philadelphia, fifteen hundred at Maidenhead, ten thousand back in Philadelphia, and so forth—Whitefield estimated that hundreds and thousands of people attended his sermons.[41] Thus all through the month of November Pennsylvania, New Jersey, and Delaware rocked to the revivalism of Whitefield. The following April Whitefield again toured the middle colonies and enjoyed the same enthusiastic audiences.

Whitefield frequently preached outdoors, like the conventiclers of seventeenth-century Scotland, either from balconies overhanging city streets or in fields. Due, in part, to the vast numbers of listeners, too many to be contained in a church building, the field seemed to Whitefield to be the preferred location for preaching. In England "the generality of people think a sermon cannot be preached well without; here they do not like it so well within the church walls."[42] He always spoke in the streets of Philadelphia—his followers built him an outdoor rostrum, and toward the end of this first tour, Whitefield reported preaching to five thousand from a balcony in Chester.

> It being court-day, the Justices sent word they would defer their meeting till mine was over; and the minister of the parish, because the church would not contain the people, provided the place . . . I was told that near a thousand of the congregation came from Philadelphia.[43]

Everywhere he traveled, he reported that people wept, groaned, and suffered under convictions. Even that great skeptic of the age, Benjamin Franklin, could not help but be impressed with

> the extraordinary influence of his oratory in his hearers, and how much they admired and respected him, notwithstanding his common abuse of them, by assuring them they were naturally half beasts and half devils.

Certainly no one was more surprised than Franklin at finding himself subject to the persuasive oratory of Whitefield. Although Franklin absolutely disapproved of Whitefield's plan for an orphanage in Georgia, Whitefield had "finish'd so admirably, that I empty'd my pockets wholly into the collector's dish, gold and all."[44] Among the most graphic descriptions was Whitefield's account of Samuel Blair's congregation of Faggs Manor:

> The bitter cries and groans pierce the hardest heart. Some of the people were as pale as death; others were wringing their hands; others lying on the ground; others sinking into the arms of friends; and most lifting up their eyes to Heaven and crying to God for mercy.[45]

Although Whitefield drew enthusiastic responses wherever he went, the revival often disappeared as quickly as it had come. Many of his listeners were skeptics, like Franklin, and while Whitefield often noted the number who came to scoff and stayed to repent, no concrete evidence supports his impressions. Directly after Whitefield's departure in 1740, the revivalism continued in only two areas: New England, awakened at least five years before by Jonathan Edwards, and the middle colonies, where Presbyterian clergy had been nourishing revivals for the previous ten years. Because Whitefield was such a gifted preacher and because he was an itinerant, he was able to reach large numbers of people who had heretofore been untouched by the Spirit. These people were the unchurched, but not in the sense that they never had attended or belonged to a church. Rather, they were individuals familiar with religious patterns who were currently unserved by the existing institutions. Whitefield's activities rekindled their interest in religion, reminded them of past traditions. Many of these middle-colony communities established congregations and, upon Whitefield's or another itinerant's advice, would ask a presbytery for ministerial support. From this perspective, George Whitefield did not initiate significant cultural change in the colonies. He did, however, reinforce and escalate trends that were traditional to Scots-Irish culture.

Only recently have historians begun to see George Whitefield as part of a larger movement, rather than as the final cause of the awakening. For years scholars have asked why Whitefield became the focus of a primary religious movement. William Kenny describes Whitefield as a Weberian charismatic leader who appealed to the inclinations toward dissent popular in the new pluralistic society. Martin Lodge also lays out the religious circumstances of the eighteenth-century middle colonies as a heterogeneous arena uncontrollable by Old World churches. Both perceptions assume no continuity between old patterns and new, as if the home communities had provided no models upon which to build. For the Presbyterians, the Ulster experience had provided good training for their position as a dissenting minority. Moreover, the religiosity favored in Ulster and much of Scotland prepared the Scots-Irish to respond to Whitefield. George Whitefield may have introduced new techniques to the English, and he may have raised revivalist emotionalism to new heights of enthusiasm. Nevertheless, among the middle-colony Presbyterians was a large community who recognized, understood, and favored Whitefield's approach.[46]

This conception of George Whitefield as a catalyst rather than an instigator is amply demonstrated by his experience in Scotland. As Whitefield traveled throughout Scotland, east, central, and west, he experienced his greatest success among two groups. Several congregations and clergymen in west Scotland, traditional revivalist territory, were wildly enthusiastic for revivalism. Because many ministers in this area supported the revivals, they lasted for several years. The other community seriously affected by Whitefield was the Secession Church, again an organization already committed to enthusiastic rituals. In the case of western Scotland, Whitefield effectively revitalized a passing tradition. Among the Seceders he provided new meanings for old rituals. In fact, one historian has argued that only the revivalist option offered by Whitefield and accepted by many clerics of the evangelical party could have kept several congregations out of the Secession Church.[47] In neither Scotland nor the colonies did Whitefield start a new religiosity; he reinforced and encouraged the old. That he enjoyed a greater, more enduring success in the colonies indicates that while Whitefield was an important catalyst, the size and scope of the awakening's revivalism depended upon the strength of the traditions and the support given it by others.

An important aspect of Whitefield's presence in the colonies that should not be overlooked is the focus he provided for the organizers of the opposition. Whitefield's theology, from a Presbyterian perspective, was unsound, if not heretical. He openly, verbally encouraged emotional responses to his sermons and evinced a toleration of denominational differences with which Presbyterians were not entirely comfortable. He was very young, only twenty-five during his first two tours, and he had focused his energy upon experimental piety. Thus, his personal theology was not fully developed.

When he arrived in Philadelphia, Whitefield was closely following the Arminian philosophy of Wesley; after his initial month touring with the Tennents he became convinced of the truth of Calvinist Christianity and the sanctity of Presbyterian government. Because Whitefield preached and published before and during this transition, his orthodoxy was open to attack from the Old Lights. An anonymous pamphlet entitled *The Querist* appeared in 1740, citing and challenging several extracts from Whitefield's writings, extracts which were, from a Calvinist perspective, highly questionable, particularly references to universal salvation and the efficacy of sincere repentance. Attributed by Samuel Blair to David Evans, *The Querist* included thirty pages of intelligent, fault-finding argument. Since the points raised were unanswerable, Whitefield did not try. Rather, he further aggravated the Old Lights by publishing a brief, eight-page apology for unguarded expressions, attributing his errors to his youth, and attacking *The Querist*'s author for lack of charity. A few more pamphlets were exchanged, but since most of the divisions had already been suffered, these publications accomplished little except added hostility on both sides.[48]

When George Whitefield rejected an emphasis upon intellectually correct theology in favor of experimental piety, when he, in fact, deprecated the criticisms raised by his opponents as of secondary importance to their failure to encourage "spiritual vitality," he thoroughly alienated a group already suspicious of the revival. George Gillespie, who had been undecided during the early stages of the revival, now portrayed Whitefield as "a Man under a delusion of Spirit . . . a Strengthener of erroneous Persons, as Arminians, Moravians, and a Puller-down of a Reformation Work."[49] The laity, on the other hand, did not participate in these intellectual debates at all, displaying an obvious preference for the "spiritual vitality." The New Light ministers might have been caught in the middle had not Whitefield admitted his errors and praised the Westminster Confession. However, since he had recognized his errors and declared that he no longer held such heterodox beliefs, revivalists felt free to reaffirm the primacy of true conversion over right education. Between Whitefield's first and second visits, Tennent delivered his *Danger of an Unconverted Ministry*, and during this time complaints about intrusions rose to dangerous levels. By 1741 the foundation had been laid for schism.

The Synod of 1741 was almost a direct response to the turmoil of the previous year. The New Lights had managed in 1740 to overturn the ruling against itineracy. "The Synod do now declare yt they never tho't of opposing, but do heartily rejoice in the Labours of ye Ministry in other Places besides their own particular Charge." The synod, in fact, ruled that all clergymen would henceforth behave as if the overtures of the previous years had never been passed. In the minutes, this statement directly preceded the qualification of their ruling on presbyterial authority to ordain.[50] These two resolutions, combined with Blair's and Tennent's paper on ministerial defects, must have frightened the Old Lights, for the next year they came prepared for battle.

One month before the synod met, in April 1741, the Presbytery of Donegal considered an overture proposed by John Thomson and his elders. Citing the recent discord in congregations whose people "are So exceedingly amused & captivated with ye Shew of extraordinary zeal & piety," the presbytery lamented the "divisive, uncharitable rash Judging, [and] disorderly practice" of those ministers perceived as fomenting the divisions. The presbytery passed a set of resolutions that restricted admittance to the Lord's Supper to those who personally subscribed the Westminster Confession and Catechisms and who promised to submit to Presbyterian government as laid out in the Westminster Directory. Moreover, each communicant promised "not to countenance or encourage these disorderly preachers that have been invading our Congns. in a disorderly way by going to hear them especially these who are under any prohibition by act of Synod of Pby." Finally, no minister was to allow "Such preachers to preach in their pulpits or Congns., nor go to hear them on pain of censure." There follow several pages of justifications, which can be summarized in assertions that Donegal's behavior was biblically sound and traditionally correct, and in their hope that they could protect the poor, ignorant people from being deluded by those self-aggrandizing preachers.[51] After such a successful handling of the intrusions in Donegal, it perhaps was not a large step toward the same sort of action in the synod. On the fifth day of the Synod of 1741, Robert Cross brought in a protestation against the New Brunswick Presbytery, and the Log College men were summarily cast out of the synod.

Gilbert Tennent's belief that this course of action had been well planned had much support, despite the spontaneity recorded in the minutes. The Synod of 1741 opened with the absence of the entire New York

Presbytery. This presbytery habitually stayed away from synod meetings whenever trouble was expected, and the delegates later admitted that, fearing an open confrontation, they had consciously kept away in order to be in a better position to effect a reconciliation. Then, after the opening sermon, the first order of business was not the standard appointment of the synod committee, but the consideration of Alexander Creaghead's behavior. The afternoon business session was occupied with charges brought against John Thomson by the congregation at Middle Octara, as well as with Creaghead's case. For three days the synod business of absences, excuses, and collections was continued, while the appointment of a synod committee, usually first on the agenda, was continually deferred. When Cross's protest was finally produced, the minutes record that the New Brunswick Presbytery, discovering itself in the minority, quietly withdrew from the synod (an action singularly inconsistent with the New Lights' usual, noticeable, public practices).

Gilbert Tennent later wrote that the New Brunswick brethren were not a minority in opposition to the protestors, for while twelve ministers signed with Cross against the ten Log College ministers, neither George Gillespie, Jedidiah Andrews, Alexander Hutchison, nor Francis McHenry chose either side. Moreover, Tennent recalled a general confusion that ensued while several ministers rushed forward to sign the protest. Then business proceeded, including the appointment of a synod committee that included no New Brunswick members. Additionally, Tennent noted, he and his colleagues did not "withdraw" until the synod meeting was over. Still, by the nature of the protestation, the members of the New Brunswick Presbytery were cast out of the synod.[52]

The protestation itself began with an invocation of the Westminster Confession, thus linking the issues of 1741 with those of 1729, a continuity that was absolutely rejected by the New Brunswick members. After declaring their conscientious inability to sit in synod with the others, the protestors proceeded to give their reasons: the intrusions, the New Brunswick position on the examination of candidates, and the references to some individuals as unconverted ministers. The protesters also complained of New Brunswick's allowance that persons with conscientious scruples need not abide by the rulings of the higher judicatures. Most of the vehemence, however, was saved for the lay response to the intruding itinerants, who had continued

> in their disorderly Itinerations, & Preaching through our Congregations, by which (alas! for it) most of our Congregations thro' weakness and Credulity, are so shatter'd and divided, and shaken in their Principles, yt few or none of us can say, we enjoy the Comfort, or have the Success among our People which otherwise we might, and which we enjoy'd heretofore.

The people had begun to "judge their Ministers to be graceless, and forsake their Ministers as hurtful rather than profitable." Nor did the protestors neglect the method of the revivalists.

> Their Preaching the Terrors of the Law in such a Manner and Dialect as has no Precedent in the word of God, but rather appears to be borrowed from a worse Dialect; and so industriously working on the Passions and affections of weak Minds, as to cause them to cry out in a hideous Manner, and fall down in Convulsion-like fitts, to the marring of the Profiting both of themselves and others . . . and then after all, boasting of these Things as the work of God.[53]

John Thomson continued the hostilities by publishing a polemic against the actions of the New Brunswick Presbytery of the previous three years. He addressed Tennent's protest of 1739 and his and Blair's papers of 1740, as well as a variety of unofficial activities. Thomson called the revivals "these disorderly violent new-fangled Notions and Stirs about Religion."[54] Gilbert Tennent responded with equal vehemence, deprecating if not openly mocking the work of his opponents: "As to Comfort, we believe them, but respecting Success, we thought it had been the same as formerly; for, truly, this is the first we have heard of the Success of most of them."[55] The Presbytery of New York had waited too long to interfere.

The New York ministers at the Synod of 1742 attempted a reconciliation under mutually acceptable arbitrators, but the protestors refused to allow the absent members any authority in defining procedure. In response, the New York ministers, supported by Francis Alison of Philadelphia, protested the exclusion of New Brunswick as illegal procedure. Again in 1743, New York tried to reunite the two. "Reasonable proposals" were sent to the withdrawn brethren only to be refused until the 1741 protest was revoked. The next year the frustrated New York Presbytery declared its sympathy with the excluded ministers, and in 1745 the Synod of New York was formed out of the presbyteries of New Brunswick, New York, and half the members of New Castle. The remaining New Castle members, together with those of the presbyteries of Philadelphia and Donegal, continued as the Synod of Philadelphia.

It is tempting to perceive the schism of 1744 as the culmination of the Great Awakening among middle-colony Presbyterians, but this would confuse what was an essentially institutional crisis with the broader, popular movement. Instead, the schism should be understood as an institutional reflection, a clerical response to the religious phenomena among the people. Unlike the Irish subscription controversy of the 1720s, the official resolutions and protests that led to division in the colonial church evidenced little active lay participation at the synod level. Congregations were certainly rewarding some styles of religiosity and discountenancing others, but they did not try to bend the church organization to serve their needs. Because some ministers emphasized emotional piety and conversion, the laity were free to devote their spiritual energy to the revivals themselves. The people no longer had to fight; they merely expressed their preference. The few times the laity were seen involved in official matters, they were stopping the church's interference in the progress of the revival, insisting that the judicatures leave well enough alone. Now the people needed neither subscription nor any other symbol to reassure them that tradition was honored. In the Great Awakening they had the tradition itself.

Within the battles resulting in the schism could be seen the wide range of the ministers' reaction to the religious tension. Three separate issues have been identified as the points around which debate revolved. First was the problem of ministerial qualifications, an issue originally raised before the revivals by the arrival of Irish ordinands who were sometimes uneducated or immoral, occasionally unsound. On one side the Old Lights favored a special, academic examination for any candidate lacking a university baccalaureate, including graduates of Tennent's seminary. On the other were the New Lights, the Tennents and their followers, demanding that evidence of experimental piety be a criterion for ordination. In both cases the advocates managed to insult and undermine the authority of their opponents.

A second issue was the itineracy of the Log College men and their intrusions into unwelcoming presbyteries and pastorates. Although the New Light preachers were supported and regularly invited by residents in a particular region, they were perceived by the Old Lights as violent, hostile intruders coming to steal and alienate congregations from their rightful pastors. These perceptions were not as irrational as they first appear, for even if the New Lights had no such intention (and this is questionable), their actions often brought the feared result.

The third issue, of course, was the revivalism itself. Clerical opinion on the quality of the revival did differ, with several ministers such as Gillespie who were uninvolved in the other issues attempting to make sense of the new religiosity. Many moderates were obviously confused and open to both positions, collecting information and opinions, such as Whitefield's own theological writings, until they felt able to render a fair evaluation. With all three issues the religiosity of reason confronted the spirituality of emotion, and party lines were drawn.

John Rowland assumed an interesting symbolic role in these debates, for in Rowland all three issues came together. Here was the candidate praised for his experimental piety and licensed by the Presbytery of New

Brunswick without a formal, academic examination. Here also was a probationer who crossed the bounds of the Presbytery of Philadelphia at the invitation of the congregation and to the dismay of the presbytery. He did in fact divide the congregation and "steal" half the people away from the Old Light candidate. Here too was a master of the revivalist technique who "proclaimed the terrors of the divine law with such energy to those whose souls were already sinking under them that a few fainted away." An observer further recorded that Gilbert Tennent interrupted Rowland's address with the question "is there no balm in Gilead? is there no physician?" Rowland responded and dropped

> the terror of his address, and sought to direct to the Savior those who were overwhelmed with a sense of their guilt but before this had taken place, numbers were carried out of the church in a state of insensibility.[56]

While Rowland apparently, in the estimation of Tennent, overstressed the terrors, he was, unlike his predecessor Glendinning, able to correct his error and lead his listeners beyond the terrors of the law to the joys of salvation.

Additionally, Rowland's brief but tempestuous career reveals in microcosm the role played by the laity during these years. For example, Rowland was esteemed by his colleagues as a gifted, successful minister, for he was able to gather a large crowd and awaken the sinners. The laity, then, served as a measuring stick for the preachers. Few would claim that the choice of the people was infallible. Dickinson criticized the congregants at Woodbridge for their reception of Pierson, and Tennent himself would soon come to challenge evangelical preachers like the Moravians, despite their appeal to the people. Nevertheless, popularity enhanced a minister's reputation among his peers.

The response of the Hopewell and Maidenhead congregation is another example of lay choice and the power realized in making that choice. Although set with a minister, several members invited Rowland to preach, rejected the previous candidate John Guild, and, when reconciliation became impossible, split the congregation in two. The New Brunswick Presbytery was certainly involved in this case, but the split was brought about by a group of lay persons inspired by the revivalism.

Like most organizational struggles, the crisis in the colonial Presbyterian church was a battle for power. Whose colleagues, whose school, whose religiosity would ultimately decide the direction that the church would follow? The New Lights were first to realize that the laity provided a powerful weapon in this duel. In an obvious attempt to alienate congregations from Old Light pastors, Tennent's *Dangers of an Unconverted Minis-*

try swept lay support behind the Log College men. When Whitefield first arrived he had planned to travel straight to Georgia. Yet he was persuaded to remain for a brief while and preach, persuaded by men who knew their people, knew that Whitefield's success would only make their own party stronger. From this perspective, the laity seems a pawn manipulated to the aggrandizement of the Log College men.

However, a second power struggle was raging alongside the institutional fight. The laity was challenging the clergy for control of Presbyterianism. Through select attendance at meetings and occasional emotional displays, the laity made its position known. It divided congregations and built new churches, publicly ridiculed ministers and refused to pay pastors unsympathetic to its wants. From this perspective, the laity seems to have consciously rewarded the Log College men with its support; it used the New Light ministers to achieve the restructuring of the churches.

Perhaps the best answer to this question is that the ultimate victory belonged to neither the New Light pastors nor the laity. The ministers joined forces with a majority of the laity and brought the Presbyterian church to schism. The support of each group increased the power of the other. The next few years would demonstrate that this was an effective alliance, capable of gaining control of the church. Nonetheless, the clerical-lay competition should not be deprecated, for attempts to assume control were continually exerted by both groups. For a brief while, the objectives of one clerical party could be accomplished with the assistance of lay persons working toward their own goals. Within two decades, however, their goals would diverge, and the basic struggle for control would begin again.

7

Triumph of the Laity

Although efforts to reconcile the two factions lasted four years, the Old and New Light Presbyterians were, after the actions of 1741, set free from each other. The antirevivalists continued to promote their own brand of piety, characterized by theological orthodoxy, moral precepts, and holiness. The New Lights, freed from the institutional interference of their opponents, were able to drop their animosity and concentrate upon the revival itself. Previously united in defense of true religion, the laity and supportive ministers were now able to band together for the propagation of the gospel religion. Whether they would take advantage of this opportunity was another question. In other words, were the clergy and laity now ready to pursue their common objectives, or did they simply transform the battle for control into a laity-clergy struggle?

As the 1740s and 50s progressed, clerical-lay cooperation, though frequent and successful, could not always be guaranteed. Throughout the years following the schism, congregations fought with their pastors. Most often the prorevivalist clergy would support the people in their efforts against an Old Light minister, but the laity occasionally and victoriously challenged a pastor sympathetic toward the revival. On the other side, the clergy, almost as soon as Whitefield had departed, began to back away from its previously unqualified support for enthusiastic piety. These oppositional movements nourished a distrust between clergy and laity that might have led to a revealing competition. However, the affairs of the Old Light organization had progressed so badly that scarcely ten years after the official schism they were seeking reconciliation. With the reunion of 1758, the clergy-laity dynamics changed once again, with the laity needing the support of the clergy to maintain their traditional religiosity.

In order to evaluate the goals, program, and strength of the Scots-Irish laity, then, it is imperative to explore in some depth the progress of the

Great Awakening in the years following the schism. Congregations capitalized upon the revival, manipulating the extraordinary circumstances to their own benefit. Ministers increased their power through the support of the laity, only to find themselves in conflict with their own followers. Without the open cooperation of a majority of ministers, the people would have found themselves without an entry into the power structure of the church. However, the people may not have required a Presbyterian institution for the expression of their piety. Thus the struggles between the laity and the clergy during these years were also reflective of the relationship between popular religiosity and the institutionalized church.[1]

Only seventeen years after the New Brunswick brethren were cast out of the synod, the leaders of the Old and New Light parties reached an agreement, and the two were reunited into the Synod of New York and Philadelphia. Considering the brief duration of this separation, the ultimate significance of the schism in the overall history of the colonial Presbyterian church is difficult to ascertain. On one side, the composition of the clerical and lay communities indicates that the church of 1765 would have been the same with or without the schism. On the other, those same communities pushed the church toward separation. Moreover, in the 1740s, this division held great symbolic import for both sides. If the New Lights truly believed that their opponents were unconverted sinners, then their withdrawal from the Synod of Philadelphia represented a rejection of the body corrupt in favor of a purified community. So too, since the Old Lights affirmed that Tennent and his followers operated under a delusion, the removal of the New Brunswick Presbytery was a grand purge. Perhaps the best method for evaluating the significance of the schism, one fair to both retrospective and contemporary analyses, would be to approach the schism as a symptom of eighteenth-century religiosity, rather than as a causal factor in the church's development.

Whatever the initial intentions of the schismatics, the end result was a divided church, two churches able to develop and function without major institutional opposition. The antirevivalists tended to remain tied to their opponents, requiring the revivalist "enthusiasm" as a foil against which their own brand of Presbyterianism appeared orthodox, moral, and reasonable. Even the titles of the sermons show this continued preoccupation. John Thomson's *Doctrine of Conviction set in a clear Light*, for example, was produced by a man who felt the need to prove, perhaps to himself, but certainly to his congregants, that his quiet, nonemotional piety was not merely an acceptable alternative, but the true course of religion in opposition to a whimsical, frenzied style of religiosity that focused on convictions and convulsions. He constantly attacked the principle of rash judging, doubtlessly in defense against the condemnations that he believed were raised against him.[2] Except for this sort of polemical literature, the Old Lights published very little as their ministers declined in reputation and influence.

On the other hand, the New Lights left their opponents behind and watched awestruck as the Spirit moved. The promises of the 1730s revivalism had finally burst forth as the Great Awakening swept through Presbyterian areas. Jonathan Dickinson, one who had, in 1739, held himself aloof from the revival as he observed the behavior of Pierson's congregation at Woodbridge, gave his support to the revival as early as 1742.

> The most serious and judicious, both Ministers and Christians, have look'd upon it to be, *in the main*, a genuine Work of God, and the Effect of that Effusion of the SPIRIT of Grace, which the faithful have been praying, hoping, longing and waiting for. . . .

Like the apologists for the Six-Mile-Water Revival, Dickinson accepted the emotional, physical rituals with reservations: ". . . some Circumstances attending it, [are] from natural Temper, human Weakness, or the Subtlety and Malice of Satan permitted to counter-act this Divine Operation." Thus Dickinson emphasized the less emotional features. For example, most of the awakened sinners did not cry out, and those who were converted displayed a reformation of behavior. He could not, obviously, escape the accusations of enthusiasm, and the majority of his justifications of the revival were laid out in the plain-style fashion of objections and answers revolving around those physical manifestations. Of course listeners were frightened and terrified; all sinners should fear their just punishment. Dickinson was following a long descent of Calvinists, including the seventeenth-century John Cotton and eighteenth-century Ebenezer Erskine, when he distinguished between gospel light and doctrinal, or legal, knowledge. People must not only have an intellectual understanding of their sinfulness, they must also be sensible of it. "How can you depend upon Christ to save you from the Wrath to come, when you have no realizing Apprehension of your being expos'd to that tremendous Wrath?"[3]

Dickinson was concerned about the spiritual pride arising out of the revival, but he had answers for that as well. He admitted that the converted talked about their conversions, but asserted that religion ought to be the main topic of conversation. He was far more anxious about the practice of judging others, what he called a "sinful censoriousness," aimed not only at profligate sinners, but also at otherwise blameless individuals merely because they had not experienced the revival's enthusiasm. He was especially irate, perhaps out of personal experience at Woodbridge, with those who presumed to judge "those *Ministers* of the Gospel, who are visibly *well-qualified* for the Ministry, and have visibly *conducted* themselves *well* in the Discharge of their sacred Trust. . . ." However, he added, no one, even the converted saint, was perfected until death. He, like Robert Blair, worried about false conversions and took pains to weed out the hypocrites. Of tremblings, swoonings, and convulsions he had heard rumors, but he claimed never to have directly observed such behavior. He had witnessed the raptures of the saved, but such expressions of joy should surprise no one. The Lord worked in mysterious ways, and Dickinson intended to suspend judgment until he actually observed the enthusiastic behavior.[4]

Samuel Blair recorded the spread of the revival to his congregation at Faggs Manor, Pennsylvania in 1740 with the visit of Alexander Creaghead. Blair's congregation had had a clear predilection toward revivalism, as evidenced by the application of some members to the Associate Presbytery in Scotland and their favorable response to Creaghead. The revival began at Faggs Manor with a few people bursting into tears during Creaghead's preaching. Upon the resumption of his pastoral duties, Blair received the same sort of response, which he claimed he tried to discourage. Like Dickinson, Blair was careful to discuss the reformation of behavior, and he particularly noted the conversions of several notorious sinners. Of special interest was Blair's report that the "sacramental Solemnities for communicating in the Lord's Supper, have generally been very Blessed Seasons of enlivening and enlargement to the people of God." He further noted that many experienced ecstasies during Sunday communion services and the Monday thanksgiving rituals. Blair cited the case of one woman, anguishing over the state of her soul, who saw her first hope of reconciliation and mercy on a Sunday evening after a communion service. After experiencing her final conversion, she would speak of "unspeakable Ravishments of her Soul at a Communion Table."[5] Although provocative oratory had become a necessary part of the revivalist ritual, the centrality of the sacraments had not been replaced. In fact they were used, as in the seventeenth century, to further conversion.

The overwhelming support of the laity for the revival was unquestionable. The varying growth rates of the synods spoke for the New Lights' popularity. During the thirteen years that followed the 1744 rupture, the number of Old Light ministers declined by six, or 23 percent, while the New Light ranks gained forty-four members, a 157 percent increase. The Synod of New York now trained native-born ministers for its service in the newly established (1746) College of New Jersey, and it might be argued that as the primary purpose of the church was service to congregations, the ability of the Synod of New York to supply more ministers could have accounted for its popularity. Yet in those localities where an Old Light cleric was installed, New Light congregations often split off at their own volition, requesting independently supply from a New Light presbytery.[6]

In the Philadelphia Presbytery, not only did the Hopewell and Maidenhead congregation split in two, but so did the congregations of Cohanzy, Neshaminy, and Great Valley. In Philadelphia, following Whitefield's abundant success in the city, people of several denominations gathered together to pursue the work of the Spirit. An early, anonymous historian counted among Whitefield's followers Quakers, Anglicans, Congregationalists, some German Lutherans and Pietists, some unchurched, and a plurality of Presbyterians, despite the adherence of the First Presbyterian Church to traditional doctrine and liturgy.[7] At first the Presbytery of New Brunswick simply supplied the New Light adherents with preaching. The people organized themselves into a congregation and called Whitefield to be their pastor. Perceiving his vocation as primarily itinerant preaching, Whitefield naturally refused, though he recommended his friend Gilbert Tennent as one of the staunchest supporters of his work. Although Tennent was unwilling to leave his charge at New Brunswick, he and his presbytery were heavily influenced by "the dangerous and difficult Scituation" of the people of Philadelphia, as well as a scarcity of candidates. In August of 1743, Gilbert Tennent accepted the call.[8]

In 1744 the Second Presbyterian Church of Philadelphia was officially recognized as a congregation by the Synod of New York. The official membership was approximately two hundred adult members. After devoting five years to building up its resources, the congregation began planning actively for its own building. After acquiring a lot, Tennent and his lay advisors estimated the cost of a simple but adequate building at two thousand pounds, of which only a third could be provided by the congregants. Using his masterful rhetorical abilities, Tennent convinced many Philadelphians who were not church members and many New Light colonists who were not Philadelphians to subscribe to the project. By the end of his campaign, men like Governor Reed and William Allen, a leading congregant of the major competitor—First Presbyterian, had made sizeable contributions.[9] The church, easily financed, was quickly built, and less than two years after the first subscription was sought the new congregation moved into the building from which it would dominate the Philadelphia Presbyterian community. During the first twenty years the church's governing board authorized several additions to a church edifice that already accommodated 240 families, including an addition of pews in the east gallery, construction of a second gallery and a steeple gallery, and the placement of pews down the center aisle.[10] The very fact that such additions were deemed necessary, and that they were paid for, illustrates the strength of this community.

The Presbytery of New Brunswick saw much congregational activity as one church after another asked to come under its jurisdiction. In 1740 the congregations from Tinnacum and Newtown asked to be dismissed from the Presbytery of Philadelphia and joined to New Brunswick, a request that was granted. The people of Tredyffrin, without a minister for several months, in 1741 asked the Presbytery of Philadelphia to dismiss them to the care of New Brunswick.[11] Four congregations in New Castle Presbytery placed themselves under the care of New Brunswick, while five communities—New Castle, Drawyers, Red Clay, Elk River, and Pencader—divided. And in Donegal, almost every congregation either withdrew as a unit from the presbytery or suffered a separation.

As early as 1740 the congregation at Nottingham, led by elders Abraham Scott and John Kirkpatrick, asked to be removed from the jurisdiction of Donegal.[12] In August of 1741, only three months after the New Light contingent had withdrawn from synod, Adam Boyd reported that most of his congregation had left him for the sake of the New Brunswick preachers. He affirmed that he would continue to serve those church members who remained, but he requested a second charge for financial as well as occupational/emotional reasons.[13] By June 1743 William Bertram reported that his congregation had separated. Several members had brought in a complaint against Bertram and his elders, the Londonderry Session, for admitting John Ireland as an elder. Ireland was accused of criticizing, insulting, and harassing people who attended the field preaching. Ireland had admitted that his expression was "unguarded" and had therefore been absolved by the session. The presbytery approved this action, doubtlessly forfeiting a few more Derry congregants to the new church.[14]

Even John Thomson, stalwart of the anti-revivalist cause, had quite a difficult time with his congregation at Chestnut Hill. He had taken the lead in the battle against the Tennents, not only publishing pamphlets, but proposing that Donegal assume a leadership role in this struggle. With the support of his elders, Thomson forged ahead with his animadversions. At the April presbytery meeting of 1741, Thomson and his elders presented the overture, which passed, to hinder people and clergy from encouraging disorderly itinerants. At this same meeting someone presented a petition, in the name of the congregation, requesting Thomson's demission. This document was proven to be a forgery. Scarcely a month later, in full synod, Alexander Creaghead's congregation at Middle Octara brought charges against Thomson. These charges were never prosecuted due to the synodical division at this meeting, but Thomson's problems did not end with the division. For the next three years he had increasing difficulty getting his salary until, in August 1744, Thomson requested and was granted a demission.[15]

On the other hand, congregations of New Light ministers generally experienced no defections, though this should not be interpreted as indicating no problems. The Suffolk Presbytery, created in 1747 to gather the new Long Island congregations under a judicature, frequently dealt with disagreements among pastors and various lay members. Nathaniel Mather's congregations at Aquebogue and Jamesport were not certain that they wanted to join with the Presbyterians. "As some of Mr. Mather's Chh. and Congregation had turn'd Separates, so others appeared to have a List that way." So too the Bridge-Hampton church was divided over organizational loyalty. The pastor, Ebenezer White, had to repent publicly several "rash comments" before the process of reconciliation could even begin. Of course, since these battles involved congregational versus presbyterian polity, it may be that these were ethnic divisions. Yet whatever the causes of such divisions, it must be noted that the Suffolk Presbytery supported pastors and reconciliation, and was fairly succesful at both.[16]

Even New Brunswick had its share of problems. Andrew Hunter, who had responsibility for two congregations, began to have problems with Deerfield in 1753. The problems lasted until 1760, when Deerfield and Greenwich were permitted to separate.[17] And there was the peculiar argument of the pastorless Tehicken congregation over the location of its meetinghouse. The congregation had decided to build a new meetinghouse, and since the membership was spread out geographically, it had decided to choose the location by lot. One faction in the congregation now complained of the decision and asked that the lottery be ignored. Neither the presbytery nor the synod (to whom the matter was eventually referred) approved the use of lots. However, since the lottery was drawn fairly, and since congregants had agreed to bind themselves to the lottery, both presbytery and synod judged that they were bound. Then the complainants asked that preachers be supplied to the old meetinghouse, a request that was declared un-Christian and unanimously rejected by both judicatures. That half of the congregation even asked to leave the jurisdiction of New Brunswick, and this request was flatly refused. Yet peace was eventually achieved, for the congregation sought baptism for the children and, in order to obtain the services of a minister, agreed to abide by presbytery rule.[18]

Undoubtedly these were aggravating problems: a congregation selecting a meetinghouse site by lot, people who could not afford or refused to pay their ministers. Nevertheless, these were not the same sort of problems as those suffered by the Old Light party. The laity, when frustrated, fought for their rights, but at this point they did not leave the church. They might threaten to leave the Presbytery of New Brunswick for Abington, but none were leaving the Synod of New York. In order to obtain the sacraments, the people at Tehicken were willing to put aside their differences, at least temporarily. The fact remains that while the Old Light ministers struggled to hang on to their pastorates, the New Lights had more than they could handle. The June following the 1741 Protest saw sixteen congregations ask the Presbytery of New Brunswick to take them under its care.[19]

The awakening was spreading in other directions as well. When Whitefield toured the colonies in 1740, he had stopped at the capital in Virginia, though no one from Hanover came to hear him preach. Three years later a young Scot who had been visiting his home brought from Scotland a book of sermons Whitefield had preached in Glasgow. A lay person read those sermons publicly, and several persons were awakened. As they found themselves to be Presbyterian in theology, the people of Hanover asked the Presbytery of New Brunswick to supply them with a minister, and William Robinson was sent for a brief tour. Robinson preached four days to larger and larger field assemblies. Because his stay was limited, Robinson helped the people establish regular worship services using lay readers. Hanover remained under the care of the Synod of New York, with the community growing until five meetinghouses were constructed. The first Presbyterian minister to preach from these pulpits was John Blair. Although the people had requested a permanent leader, Blair could only remain a short while. Yet during his visit, it seemed the spirit had revived.

> One Night in particular a whole House-ful of People was quite overcome with the Power of the Word, particularly of one pungent Sentence that dropt from his Lips; and they could hardly sit or stand or keep their Passions under any proper Restraints. . . .[20]

Samuel Davies, a graduate of Samuel Blair's academy at Faggs Manor, arrived in Hanover in 1747. After a few months' work he had to rest and recover from consumption, but in April 1748 Davies returned to take up permanent residence among the Hanover congregations. Within six months he was dividing his time among seven meetinghouses. For each congregation of fifteen or twenty families, he would be providing preaching to four or five hundred hearers. By 1751 Davies reported three hundred communicants; at one service he baptized forty adults. The interest in religion had spread through several western counties of Virginia, including Frederick,

Augusta, and Lunenberg. The revival had moved as far south as North Carolina, where a new congregation was seeking ministers, and west to a settlement along the Susquehanna River. Even Somerset County, the area that had been assigned to the failed presbytery Snow Hill, experienced revival.

Perhaps the best indicator of the success of religion in these parts was the work of John Thomson. Thomson had left his uncomfortable charge at Chestnut Hill to do missionary work in the back country of Virginia. No longer feeling threatened by the homiletic accusations of the New Light ministers, Thomson could appreciate the work and progress of the revival in Hanover. Although he never left the Old Light Synod of Philadelphia, Thomson devoted his time, energy, and rhetorical gifts to the revival in western Virginia, where he became a very successful evangelical until his death in 1753.[21]

Revivalism had also moved northward, and as late as 1764 enthusiasm was breaking out on Long Island. Contemporary observers of these and other revivals continued to emphasize the permanency of the conversions over the emotional expressions, but most writers generally admitted the presence of enthusiastic outbursts. Samuel Buell hesitated to describe what he labeled "unusual symptoms," though he did admit that several members in his congregation indulged in shouting and weeping. His respondents were reported to have experienced the standard conversion, and he proudly asserted that at one point ninety-eight adults joined the church.[22]

Throughout all these pastoral reports on the spread of the awakening, the sacraments continued to play a central role, both as a measuring stick of success and as part of the conversion ritual. Samuel Blair's emphasis upon the celebration of the Lord's Supper in his congregation has already been noted. Davies, too, discussed some glorious communions. Moreover, he counted converts not according to professions of faith, but by communicants and baptisms. Samuel Buell described a Lord's Supper held in East Hampton that included a time of trial so that people could prepare for the sacrament and test their worthiness. Not only did ninety-eight individuals join the church beforehand, but twenty-four persons experienced conversion during the trial days of the service. Even a congregation angry at the judicatures backed down in order to obtain baptism or communion. Far from moving toward a singular emphasis on preaching, the Presbyterians at this point retained their sacramental focus.[23]

With the support of the laity, then, the New Light ministers secured their dominance of the middle-colony Presbyterian community. The New Light congregations were generally larger than the original parishes. In this attraction of the laity, there was evidence of both defectors from other churches, such as Anglican, and converts among the unchurched. Nevertheless, the fact that so many of these new, unchurched areas were locations of recent immigrant settlements indicates the prevalence among the enthusiastic laity of the Scots-Irish. Samuel Davies noted specifically that his followers tended to be Irish; they had initially settled in New Jersey and Pennsylvania and had moved south. The only group large enough to undermine the dominance of the Scots-Irish would have been the English Puritans. Yet even in the well-established communities of northern New Jersey, the Scots and Scots-Irish were beginning to challenge the New Englanders.

Historians have long debated the cultural background of the Presbyterian ministers who participated in the Great Awakening. Leonard Trinterud only followed tradition when he positioned the Old Light, "Scots-Irish" party against the "New England" and "Log College" factions.[24] If, however, the Log College were eliminated as a separate classification and its graduates were also identified by their ethnic background, the concept of the conservative Scots-Irish party would no longer make sense. The classification of all the ministers who, between 1741 and 1745, actually joined one of the schismatic synods has been summarized in Table 1. The numbers demonstrate that beyond an inclination among New Englanders to join the Synod of New York, no cultural group displayed a preference.

Similar figures for 1758 in Table 2[25] reveal the same sort of ethnic divisions. The Irish ministers were again divided, and the New Englanders continued to prefer the Synod of New York. However, all seven of the new Scottish immigrant clerics had joined the New Light synod, so that Trinterud's explanatory scheme assigning the Scots and Irishmen to the Old Light party has been contradicted by the entire contingent of Scottish arrivals. An important difference lies in the figures for native-born minis-

Table 1 Origins of Presbyterian Ministers Involved in the Great Awakening: At Schism

	Members Between 1741 and 1745		
Cultural Background	Old Light	New Light	Total
Northern Ireland	16	13	29
Scotland	2	1	3
New England	4	11	15
Wales	3	2	5
England	0	1	1
Unknown	1	0	1
Total	26	28	54

Table 2 Origins of Presbyterian Ministers Involved in the Great Awakening: At Reunion, 1758

Cultural Background	Old Light	(New Since 1745)	New Light	(New Since 1745)	Total
Northern Ireland	12	(4)	15	(6)	27
Scotland	1	(0)	7	(7)	8
New England	3	(1)	24	(19)	27
Middle Colonies	2	(2)	10	(10)	12
Other/Unknown	2	(2)	16	(16)	18
Total	20	(9)	72	(58)	92

ters. Of the twelve new ministers native to the middle colonies, ten joined the New Light synod, and from a total of thirty-two colonial-born Presbyterians ordained between 1745 and 1758, twenty-nine joined the Synod of New York. Again, what these numbers most clearly illustrate is the total inability of the Synod of Philadelphia to attract (out of the laity) and train ministers. With these low numbers and the lack of lay support, capitulation to the New Light synod was almost inevitable.

Since the Synod of New York sought union as early as 1747, it was only a matter of time, eleven years, until union was achieved. As the numbers predicted, the reconciliation agreement, or Plan of Union of 1758, heavily favored the Synod of New York, although a few concessions were made to appease the members of the Synod of Philadelphia. The first article restated the unanimous approval of the Westminster Confession "as an orthodox & excellent System of Christian Doctrine, founded on the word of God," and that the synod would continue to accept the Westminster Confession as the confession of the church. The church would also adhere to the plan of worship, government, and discipline found in the Westminster Directory. Active concurrence or passive submission to every synodical decision was required, with the clarification that this referred only to matters judged "indispensable to doctrine or government." Protests were always permitted and would always be recorded, and no member would be liable to prosecution on account of his protesting any decision (a direct reference to the right of the New Brunswick Presbytery to protest the decision concerning ministerial qualifications in 1739). Moreover, the Protestation of 1741 against the New Brunswick Presbytery was declared a legal protest, but only that, and not an act of synod. This had been the primary obstacle to any reconciliation. By acknowledging that the use of their protest as an act of synod was illegal, the Old Lights symbolically renounced all authority that they had heretofore claimed.

Concerning the two issues that actually produced the schism, ministerial qualifications and intrusions, compromise positions were reached. Probationers were to be required to produce evidence of their "experimental Acquaintance with Religion," in addition to evidence of learning and theological understanding.[26] On the other issue, ministers were warned against slandering one another with public accusations of heterodoxy, insufficiency, or immorality. Rather, according to Presbyterian discipline, anyone noticing a theological error or moral problem was first to use private admonition and, if that failed, to bring charges before a regularly constituted judicature. Moreover, a preacher was to ask the permission of the pastor, or, in the case of a vacant congregation, the presbytery, before he traveled to speak outside the bounds of his own congregation. Yet even this was weakened with the warning that "it Shall be esteem'd unbrotherly for anyone, in ordinary Circumstances, to refuse his Consent to a regular Member, when it is requested."[27]

The final article of the Plan of Union formed the new synod's official position on the Great Awakening, that a "blessed Work of God's [holy] Spirit in the [C]onversion of Numbers, was then carried on." The synod's philosophy was that "an entire Change of Heart & Life" was necessary for a person to be saved, and that "Such a Change can be effected only by the powerful Operations of the Divine Spirit. . . ." Sinners first had to be sensible of their condition and their inability to recover themselves before they might believe in Christ and his willingness and ability to save them. They had to renounce "their own Righteousness in Point of merit, depend upon his imputed Righteousness for Justification before God, and on his Wisdom & Strength for Guidance and Support. . . ." Then they might attain assurance and comfort, notwithstanding their past sins, and they would be able to continue a life of holiness. The synod did recognize that

> Persons Seeming to be under a religious Concern, imagin that they have Visions of the human Nature of Jesus Christ, or hear Voices, or See external Lights, or have Faintings of convulsive-like Fits, and on the Account of these Judge themselves to be truely converted, tho' they have not the Scriptural Characters of a Work of God above described, we believe Such Persons are under a dangerous Delusion: And we testify our utter disapprobation of Such a Delusion, wherever it attends any religious Appearances, in any Church or Time.

Still, the final vote was in favor of the awakening with the assertion that

> This is to be acknowledged a gracious Work of God, even tho' it Shou'd be attended with unusual bodily Commotions, or Some more exceptionable Circumstances, by Means of Infirmity, Temptations, or remaining Corruptions; and whenever religious Appearances are attended with the good Effects above mentioned, we desire to rejoice in and thank God for them.[28]

Although this statement represented an official affirmation of the Great Awakening as a work of God, the revival's advocates appeared to back away from a position of unilateral, indiscriminate support. Much of the emotional enthusiasm was condemned, and the awakening was accepted despite the attendant "usual Bodily commotions." Naturally, the willingness of the revivalists to challenge the more extreme enthusiastic responses rendered their position more acceptable to the Synod of Philadelphia. Yet the phrasing of this statement was not merely political compromise. Over the course of the awakening, the New Lights themselves grew anxious about the lay activities, particularly as they observed the progress of enthusiasm through the Moravian community.

Count Nicholaus von Zinzendorf arrived in the middle colonies in 1741 to oversee the development of his German-based community. His Moravians also embraced enthusiastic revival, but with a zeal that far outstripped that of the New Light Presbyterians. The Moravians would follow no pastor except one who performed expertly the revivalist rituals; whether that person had any pastoral or theological credentials was immaterial. They would desert a minister who was moral, orthodox, and even supportive of the revival if he did not actively provoke the correct response. Even more distressing to the Presbyterians was a Moravian theology that accepted universal salvation and denied the necessity of preparing for conversion. The prolonged agony of fear, terror, and despair was not considered necessary or desirable. Instead, the Holy Spirit simply swept the saved into a rapturous joy; a person remained absolutely passive throughout this experience. Moreover, the Moravians espoused the antinomian belief that once converted, a person could sin no more; a saint's very actions were sanctified.

Throughout the 1740s and 1750s the Presbyterians published criticisms and warnings against Moravian antinomianism. Samuel Blair's *Persuasive to Repentance* and Gilbert Tennent's *Account of the Principles of the Moravians* both expressed hesitance and suspicion of Moravian theology. In his *Familiar Letters to a Gentleman* Jonathan Dickinson devoted over four hundred pages of argument to enabling pastors and lay persons to withstand the attacks and persuasions of the Moravians. By emphasizing preparationism and the covenant, Dickinson satisfied the philosophical demands of God's sovereignty and the rejection of human impotence. In fact, he was in the peculiar position of defending a mandate for obedience to God's law, a position usually reserved for the Old Light ministers.[29]

The active proselytizing of the Moravians greatly disturbed Pennsylvania's religious community. Undoubtedly attractive to some Presbyterians, their efforts were generally aimed at attracting members from the German Lutheran and Reformed communities, so these were the organizations that were disrupted. However, since many among the Presbyterians shared friendships and intellectual relationships with the German ministers, they experienced vicariously the Moravians' potential to undermine church structures and clerical authority. Almost from the beginning, Gilbert Tennent refused to have anything to do with them. At one point in 1742 he wrote a fellow Presbyterian minister that his own spiritual distress combined "with the trial of the Moravians, have given me a clear view of the danger of everything which tends to enthusiasm and division in the visible church."[30] When the Second Presbyterian Church of Philadelphia was gathering, among the diverse groups was a cohesive body of Moravians. Tennent assertively informed Whitefield that he would not accept the charge of a congregation that included Moravians. Needless to say, there was no possibility of cooperation with the Moravians when that church was established.[31]

The Moravians were not the only source of enthusiastic irritation. It was difficult enough when the young, wild Presbyterian itinerant James Davenport toured New Jersey, bringing communities to uncontrolled enthusiasm. But his perambulations through New England as he provoked hysteria and burned the books of eminent Puritan divines brought shame to the middle colonies, for Davenport's original parish was on Long Island. Then Alexander Creaghead brought the anachronistic cause of the Secession Church to the colonies, demanding that all Presbyterians subscribe the Solemn League and Covenant. When the Presbytery of New Brunswick refused even to consider the documents seriously, Creaghead separated himself from the official church, leaving a large portion of his congregation at Middle Octara seeking pastoral care.[32]

However enthusiastic Tennent had been, he was too reasonable and too intelligent to miscomprehend these embarrassments and to miss the connections between the activities of those itinerants and his own. Even before the Synod of New York was organized, Tennent began to preach sermons emphasizing charity and love. In *The Danger of Spiritual Pride*, Tennent warned his listeners against Christian arrogance. He hinted that the Old Lights were guilty of this conceit when he used words such as "legal" when referring to this particular problem. Yet the characteristics that he listed: "rash-judging, back-biting . . . hatred of Reproof," as well as "the groundless Esteem and Applause of ignorant People, and the Hypocritical Panegyricks of designing Parasites" resembled the accusations raised against his own party.[33] More surprising was *The Necessity of studying to be quiet, and doing our own Business*, hardly a theme common in New Light sermons to that point. Yet here he was, enjoining people toward diligence in their duty, toward simple virtue. As he moved into the leadership of the movement working for reconciliation of the Presbyterian church, he published *Brotherly Love recommended*, a sermon originally preached to a congregation preparing to receive the Lord's Supper. In addition to the formulaic reflections on the passion of Jesus Christ, Tennent wrote of the hazards of spiritual pride, and he explicitly identified that pride with the converted.[34]

Tennent was not alone in these perceptions; his essays and sermons were mirrored by many other Presbyterian clerics. Jonathan Dickinson early warned against judging the spiritual experience of others. Samuel Davies delivered a sermon on the nature of humanity and a person's moral obligation to God.[35] As early as February 1742 Gilbert Tennent had written Dickinson of his regret for his "excessive heat of temper."[36] Whether or not the people so desired, almost before the awakening burst forward, it began to wind down in the Presbyterian church.[37]

Tennent may have regretted his temper and retracted a blanket endorsement of the awakening's enthusiasm, but he never changed his opinion that the revival was a true work of God's spirit. When the combined synods chose this view as their official position in 1758, they were not applauding the enthusiasm or ritual structure of the revival; rather they pointed to the results of the awakening: conversion. Because the participants appeared changed, because they had learned about God and had demonstrated a new, moral behavior, they were judged converted, and the revival, the instigator of those changes, was a success. This argument followed the conservative lines that most justifications of extraordinary religious phenomena followed during the eighteenth century. Conversion was a goal for all Christians, and the means by which a person could be converted was left to God. Rather than analyze the means and manifestations, those wishing to evaluate the phenomena should judge by the effects. In the end, the revivalists deserted the debates over George Whitefield, emotionalism, and physical manifestations for a firm foundation in the behavioral evidence of conversion. To repeat the synod's statement: "Whenever religious Appearances are attended with the good Effects above mentioned [change of heart and a life of holiness, among others], we desire to rejoice and thank God for them."[38]

No one questioned the need for conversion, only the methods used by the revivalists to bring it about. By 1758, however, most of the Old Light ministers had seen that, stripped of its inflammatory rhetoric, the revival preaching sought the same goals toward which they themselves had worked. When actually describing this process, the revivalists drew a model of individual conversion like the Augustinian/seventeenth-century Puritan model.[39] Conviction of sin led to terror of punishment and despair of escape. The truly saved would then see the possibility of salvation through Christ, embrace Christ as their personal savior, and come to be assured of their gracious state. Evidence of conversion would follow, with backslidings, doubts, and renewed assurances. In their use of provocative rhetoric, the revivalists had replaced the earnest study and self-examination of the seventeenth-century Puritans with emotional terrors. Preparation for the conversion received from God was not new to the New Englanders, but the openly emotional form of that preparation was an innovation difficult for some to accept, although most New England ministers in the middle colonies were, within a few years, public supporters of the revival.

The awakening's emphasis upon personal conversion could have been learned from the New England Congregationalists, whose belief in the possibility and desirability of a "visible-saint" church placed the individual conversion experience at the center of religiosity. While the New England Presbyterians rejected the visible-saint church as humane presumption and arrogance, they had nevertheless retained the conversion emphasis. For example, Solomon Stoddard did not demand that his church members display evidence of conversion; all but the notoriously sinful were permitted to participate in the sacraments. Still, the peaks of his pastoral career in Northampton were the harvests of souls. Among the Scots-Irish, however, was also a historical understanding of personal conversions that dated at least as far back as the Six-Mile-Water Revival. Certainly the narratives told by the seventeenth-century ministers and lay persons of their own conversions were very like the histories of the Great Awakening participants. Moreover, unlike the New Englanders, the Scots-Irish had a tradition, dating back to this same period, of the revivalism that fed conversions. Yet the ultimate goal sought had not been the conversion of the individual, but of the community.

Neither the Scots nor the Irish of the seventeenth century had worked toward a church of visible saints; as Presbyterians, they had adhered to a parish definition of church membership. In the beginning the purification of the congregation was attained through the conversion of its members, so that the personal/communal goal was the same. However, as the century progressed and political pressures interfered with church development, conversions of the community as a whole became the primary focus. Soon the definition of the community identity was extended to include the entire Scottish nation. After 1690, the political circumstances no longer reinforced the religious focus upon community. The ministers, restored to the leadership of the established church, switched their focus in sympathy with the Enlightenment's emphasis upon the individual. The journey toward conversion was, in the religious experience of many clerics, replaced with the mastery of rational theology, even though the laity evidenced a strong desire to continue their revivalist conversion celebrations. Additionally, during the early years of the eighteenth century, the institutional church fought to restrict the conversion rituals in the face of lay opposition.

In the colonies the centrality of conversion was restored to the Presbyterian's religious ideology by the Log College men. Whether William Tennent senior brought it from Ireland, Gilbert learned it at Yale, or he picked it up through his correspondence with Frelinghuysen, the earliest publications of the Tennents and their followers demonstrate their focus upon the personal conversion experience, rendering their religious formulations consistent with an intellectual climate that highlighted the individual. Around this center all other aspects of religiosity revolved, including their definition of the qualified minister and their revivalist practices. From the opening of the seminary at Neshaminy, experimental piety had been emphasized above all other things. The Scots-Irish laity, having never lost their revivalist tradition, supported this group of ministers as early and as strongly as possible. The laity may or may not have understood the full philosophical justification behind the revival, but within their own religious ideology, revivalism was not merely acceptable but desirable.

Was eighteenth-century revivalism centered upon the individual or the community? That the revivalist preachers aimed at individual souls cannot be denied, nor should it be forgotten that the New Englanders came from a long tradition founded upon personal conversion. During the early years of the Great Awakening in the middle-colony region there was a tendency toward congregational divisions. The inclination to separate was rather like the Congregationalists' efforts to keep out the unconverted. A community of converted members had no real need for purification. Still, the Presbyterians were not Independents, and they were not comfortable with the idea of a visible-saint church. Increasingly rigorous discipline accompanied the Great Awakening, just as it had followed the seventeenth-century revivals, and the enforcement of discipline was a communal process. The Presbyterians' continued use of the sacraments as central rituals served to reaffirm the community. Moreover, after the Old and New Lights were reunited in 1758, the clergy devoted much effort to reuniting divided congregations and presbyteries. Geographic boundaries were again invoked, in opposition to units composed of individuals with similar ideas or preferences.

Whether the ideological justification of revivalism was the conversion of the individual or the community, the very nature of revivalism was communal. Ministers frequently reported their private conferences held with congregants, affirming that such conferences brought about conversions, but the essence of the Great Awakening was in the revival meetings themselves. These were quite large, often involving thousands of participants. In the case of itinerants the meetings could last one or two days, or as long as the preacher could preach. A communion service could last three or four days, while a revival nurtured by a congregation's own pastor might involve three or four weeks of daily morning and evening preaching. The emotional response to inflammatory rhetoric followed a path set long before. As the people taught each participant the appropriate responses, revivals would generally work themselves up to a climax fairly quickly. Ministers would talk about a revival of two or three years, but they were referring to the continuous godly progress of their congregants, a perceived revitalization of piety. During these years, pastors were preoccupied with individual conversion journeys. But the actual, highly charged revival meeting that characterized the Great Awakening was an intense activity that involved the entire community for several days or weeks. In this sense alone, the Great Awakening was a great regional conversion experience.

Because the revivalism was a communal phenomenon, the critical role played by the laity cannot be overestimated. Over and over the laity proved able to control their leaders by granting or withholding emotional and/or financial support. The official options available to the laity for challenging their pastors enabled the people to rid themselves of unpopular ministers, so that the illegal retention of salary was used only as a last resort among many. As soon as the colonies had enough ministers to allow congregations some choice, congregations made their preferences clear. In their avid support of the revival and their power over the clergy, the laity did more than nourish the revival. Because they traditionally preferred that style of religiosity, the laity encouraged the majority of the clergy to embrace the revival. Just as many Irish clergy had subscribed the Westminster Confession in 1722 to appease their people, many colonial divines must have supported the revival as a protection against the loss of support.

If the Presbyterian clergy had not supported the revival, of course, the Great Awakening would not have swept the middle colonies with such speed. George Whitefield's successes, though numerous, were fleeting; usually only in congregations supported by New Light ministers did the revival flourish. A brief comparison with the Irish and Scottish situations demonstrates the importance of clerical-lay cooperation. Very few ministers in Ireland supported the revival, and throughout this same time period revivalism was peripheral to the Irish church's development. The people remained unsatisfied, and many fringe organizations split off from the General Synod of Ulster. In Scotland the picture was even plainer. Several groups of ministers supported either the Secession Church or the Whitefield movement, and surrounding those ministers were large numbers of lay supporters. Ministers who did not support the revivalism often found their churches deserted. However, because a significant number of clerics did not support the revival, church unity and strength suffered. In other words, if the church supported revivalism, the laity would rally to its banners; if the official institution did not, the laity turned apathetic.

A clergy fully supported by tithes or partially supported by a regium donum had fewer reasons to consider the preferences of the laity than did pastors recompensed entirely through voluntary contributions. Therefore, it was not surprising that the colonial clergy were more attentive to the desires of their people. Additionally, as the colonial ministers had discovered an acceptable theological rationalization for the revival, there remained no obstacle to their support of this set of rituals.

Had the clergy and laity rigidly disagreed, the awakening might have completely destroyed the colonial Presbyterian church by alienating pastors from their congregations. In Britain, the problems among the laity, evangelical ministers, and moderate pastors grew so severe that both the Irish and the Scottish churches had to be reconstructed along conservative lines in the early nineteenth century. The colonial clergy had avoided such a disaster by working with the laity; yet even during the 1740s, at the peak of the awakening in this region, there were indications of future problems. The Log College men, especially Tennent, backed away from their hard, unyielding emphasis on experimental piety toward a more comprehensive ideology that had space for charity and holiness. They completely rejected extremists such as Davenport and Zinzendorf, channeling their people toward moderation even in convictions. If the clergy did not return to the earlier patterns, there was every indication that the Presbyterian church in the middle colonies could follow a similar path of disintegration. In 1760, though, this was only a possibility.

For decades the Scots-Irish people had tried to bring their church back to the original reformation tradition. In the middle colonies in the middle of the eighteenth century, the Presbyterian clergy had finally returned to that revivalism as a hundred years of religious experience burst forth in one glorious sweeping movement. Many have laid the revivals at the feet of divines like George Whitefield and Gilbert Tennent, but clearly the clergy and laity together created the Great Awakening. In fact, in its truest sense, the Great Awakening represented neither the effectiveness of the itinerants nor this spirit of clerical-lay cooperation, but the ultimate power of the laity.

Notes

Introduction

1. Alan Heimert and Perry Miller, eds., *The Great Awakening* (New York: Bobbs-Merrill, 1967), p. xiii. For other examples, see William Warren Sweet, *The Story of Religion in America* (New York: Harper & Row, 1930); William Warren Sweet, *Religion in Colonial America* (1942; rpt. New York: Cooper Square Publishers, 1965); Richard L. Bushman, ed., *The Great Awakening: Documents on the Revival of Religion 1740–1745* (New York: Norton, 1970); Sidney Ahlstrom, "The Century of Awakening and Revolution," in his book *A Religious History of the American People* (New Haven: Yale Univ. Press, 1972), pp. 261–384.

2. Jon Butler, "Enthusiasm Described and Decried: The Great Awakening as Interpretive Fiction," *Journal of American History*, 69 (1982), 305–25.

3. Stephanie Grauman Wolf, *Urban Village: Population, Community, Family Structure in Germantown, Pennsylvania, 1683–1800* (Princeton: Princeton Univ. Press, 1976); Jon Butler, *Power, Authority, and the Origins of American Denominational Order* (Philadelphia: American Philosophical Society, 1978); William Howland Kenney, "Alexander Garden and George Whitefield: The Significance of Revivalism in South Carolina, 1738–1741," *South Carolina Historical Magazine*, 71 (1970), 1–16.

4. Edwin Scott Gaustad, *The Great Awakening in New England* (New York: Harper & Row, 1957); Alan Heimert, *Religion and the American Mind from the Great Awakening to the Revolution* (Cambridge: Harvard Univ. Press, 1966); Cedric B. Cowing, *The Great Awakening and the American Revolution: Colonial Thought in the Eighteenth Century* (Chicago: Univ. of Chicago Press, 1971); Patricia Tracy, *Jonathan Edwards, Pastor: Religion and Society in Eighteenth-Century Northampton* (New York: Hill & Wang, 1980).

5. Rhys Isaac, "Evangelical Revolt: The Nature of the Baptists' Challenge to the Traditional Order in Virginia, 1765 to 1775," *William and Mary Quarterly*, ser. 3, 31 (1974), 345–68; "Dramatizing the Ideology of Revolution: Popular Mobilization in Virginia, 1774 to 1776," *William and Mary Quarterly*, ser. 3, 33 (1976), 357–85; "Religion and Authority: Problems of the Anglican Establishment

in Virginia in the Era of the Great Awakening and the Parsons' Cause," *William and Mary Quarterly*, ser. 3, 30 (1973), 3–36; *The Transformation of Virginia, 1740–1790* (Chapel Hill: Univ. of North Carolina Press, 1982).

6. Heimert, *Religion and the American Mind*. See also Cowing, *The Great Awakening and the American Revolution*; William G. McLoughlin, "The Role of Religion in the Revolution: Liberty of Conscience and Cultural Cohesion in the New Nation," in *Essays on The American Revolution*, ed. Stephen G. Kurtz and James H. Hutson (New York: Norton, 1973), pp. 197–255; Mark Noll, "Ebenezer Devotion: Religion and Society in Revolutionary Connecticut," *Church History* 45 (1976), 293–307.

7. Harry S. Stout, "Religion, Communication, and the Ideological Origins of the American Revolution," *William and Mary Quarterly*, ser. 3, 34 (1977), 519–41; Isaac, "Dramatizing the Ideology of Revolution."

8. Nathan O. Hatch, *The Sacred Cause of Liberty: Republican Thought and Millennium in Revolutionary New England* (New Haven: Yale Univ. Press, 1977); Gary B. Nash, *The Urban Crucible: Social Change, Political Consciousness, and the Origins of the American Revolution* (Cambridge: Harvard Univ. Press, 1979).

9. Perry Miller, *Jonathan Edwards* (1949; rpt. New York: Meridian Books, 1959); Miller, "Jonathan Edwards and the Great Awakening," in his *Errand into the Wilderness* (New York: Harper & Row, 1956), pp. 153–66; Heimert, *Religion and the American Mind*; Tracy, *Jonathan Edwards, Pastor*.

10. Martin E. Lodge, "The Great Awakening in the Middle Colonies," diss., Univ. of California, Berkeley, 1964; James Walsh, "The Great Awakening in the First Congregational Church of Woodbury, Connecticut," *William and Mary Quarterly*, ser. 3, 28 (1971), 543–62; Gerald F. Moran, "Conditions of Religious Conversion in the First Society of Norwich, Connecticut, 1718–1755," *Journal of Social History*, 5 (1972), 331–43; Nash, *Urban Crucible*; Isaac, *Transformation of Virginia*.

11. Isaac's work should not be categorized with most social history, for it does not share the quantitative method. Although Isaac does not focus upon religion, as I do, in his analysis of the class components of Virginia society and culture, he is primarily concerned with symbol systems, including religion. His method is derived from Clifford Geertz, and I have found his scholarship the best model available for the historian hoping to borrow from anthropology.

12. Clifford Geertz, "Religion as a Cultural System," in *Anthropological Approaches to the Study of Religion*, ed. Michael Banton (Edinburgh: Tavistock Publications, 1966), pp. 1–46.

13. For example, the Presbyterian tradition as an adaptation to political exigencies is one theme of chapter 2.

14. Ahlstrom, "Century of Awakening," pp. 286–87.

15. Anthony F. C. Wallace, *Religion: An Anthropological View* (New York: Random House, 1966), pp. 30–39. For a practical application of revitalization theory to Iroquois society, see Anthony F. C. Wallace, *The Death and Rebirth of the Seneca* (New York: Vintage Books, 1972).

16. Perry Miller, *From Colony to Province* (Cambridge: Harvard Univ. Press, 1956).

17. William Howland Kenney, "George Whitefield and·Colonial Revivalism: The Social Sources of Charismatic Authority, 1737–1770," diss., Univ. of Pennsylvania, 1966.

Chapter 1

1. Andrew Stewart, "History of the Church of Ireland," in Patrick Adair, *A True Narrative . . . of the Presbyterian Church in Ireland (1623–1670)*, with introduction and notes by W. D. Killen (Belfast: C. Aitchison, 1866), p. 317. Andrew Stewart was born c. 1616, served the Presbyterian church during the troubled decade of the English Civil War, and died in 1662.

2. This account of the plantation of Ulster is based upon M. Perceval-Maxwell, *The Scottish Migration to Ulster during the Reign of James I* (London: Routledge & Kegan Paul, 1973). The best examination of the plantation, Perceval-Maxwell's account includes a detailed narrative of the colonization schemes, a thorough identification of the planters involved, and a quantitative analysis of the early settlers. His sources are primarily government documents including several inhabitant surveys and military lists.

3. Stewart, "History of the Church of Ireland," p. 313.

4. Perceval-Maxwell, *Migration to Ulster*, p. 279.

5. John Livingstone, *A Brief Historical Relation of the Life of Mr. John Livingstone . . . Written by Himself*, ed. Thomas Houston (Edinburgh: John Johnstone, 1848), p. 77.

6. Robert Blair, *The Life of Mr. Robert Blair . . . His Autobiography from 1593–1636*, ed. Thomas McCrie (Edinburgh: Wodrow Society, 1848), p. 57.

7. For a good survey of the early years of the Scottish Reformation, see Walter Roland Foster, *The Church Before the Covenants. The Church of Scotland, 1596–1638* (Edinburgh: Scottish Academic Press, 1975). Scottish church historians argue about the ultimate importance of John Knox to the reformation in Scotland; however to the Scots of the seventeenth century, the history of the 1560 reformation is synonymous with Knox's biography. In understanding Scottish religious culture it is not the historical reality but the strength of legend that matters.

8. William Ferguson, *Scotland's Relations with England: a Survey to 1707* (Edinburgh: John Donald,1977), p. 72.

9. Ferguson, *Scotland's Relations with England*, p. 94. Ferguson argues that from at least the twelfth century, adherence to the Church of Scotland was part of Scottish nationalism (p. 21). The connections between Presbyterianism and Scottish national identity became more pronounced in the seventeenth century and will be discussed in some detail in the next four chapters. For a detailed discussion of Scottish national identity see Arthur H. Williamson, *Scottish National Consciousness in the Age of James VI: The Apocalypse, the Union, and the Shaping of Scotland's Public Culture* (Edinburgh: John Donald, 1979).

10. Livingstone, *Brief Historical Relation*, pp. 76–77.

11. Blair, *Autobiography*, pp. 58–59.

12. A.F. Scott Pearson, as cited in John M. Barkley, *A Short History of the Presbyterian Church in Ireland* (Belfast: Presbyterian Church in Ireland, 1960), p. 6.

13. Barkley, *Presbyterian Church in Ireland*, pp. 3–4.

14. Contemporary accounts of this revival are found in the autobiographies of Blair and Livingstone, as well as in Andrew Stewart's brief history and in Patrick Adair's account. Secondary accounts of this event can be found in Robert Fleming, *The Fulfilling of the Scriptures* (1743; rpt. Charleston: Samuel Etheridge, 1806), pp. 302–4, 326–30, 333–36, and John Gillies, *Historical Collections Relating to Remarkable Periods of the Success of the Gospel* [1754] (Kelso, Scotland: John Rutherford, 1845); and, for a twentieth-century summary of previous research, see W. D. Baillie, *The Six Mile Water Revival of 1625* (Newcastle, Co. Down: Mourne Observer Press, 1976).

15. Barkley, *Presbyterian Church in Ireland*, p. 5; Stewart, "History of the Church of Ireland," p. 315.

16. Blair, *Autobiography*, p. 70.

17. Blair, *Autobiography*, p. 70.

18. Livingstone, *Brief Historical Relation*, p. 263.

19. Stewart, "History of the Church of Ireland," p. 317.

20. Blair, *Autobiography*, p. 70.

21. Blair, *Autobiography*, p. 89.

22. Blair, *Autobiography*, p. 90.

23. Livingstone, *Brief Historical Relation*, p. 82.

24. Livingstone, *Brief Historical Relation*, p. 92.BlBl.

25. *Killinchy, or, the Days of Livingstone: A tale of the Ulster Presbyterians* (Belfast: Wm. McComb, 1838), pp. 35–41, citation, p. 38.

26. Robert Wodrow, *A short account of the life of . . . D. Dickson* (Edinburgh, 1764), as cited in George Grubb, *An Ecclesiastical History of Scotland* (Edinburgh: Edmonston and Douglas, 1861), 2:339.

27. Blair, *Autobiography*, p. 19.

28. Grubb, *Ecclesiastical History*, 2:340.

29. Fleming, *Fulfilling of the Scriptures*, p. 303. Unless otherwise indicated, all emphasis in quoted matter is in original.

30. Associate Presbytery, *Act, Declaration, and Testimony for the Doctrine, Worship, Discipline and Government of the Church of Scotland* (Edinburgh: Thomas Lumisden and John Robertson, 1737). A full discussion of these events appears in chapter 4.

31. Blair, *Autobiography*, p. 63.

32. Blair, *Autobiography*, pp. 62–69.

33. The Counter-Reformation, in its effort to correct what were perceived as gross errors in practice, would reemphasize the sacraments in Roman Catholic worship, particularly the centrality of the Mass. Nevertheless, in Scotland at the

time of the Reformation the problem of noncelebration of the Mass had not been addressed by the official church.

34. Communion services could be found throughout Protestant Scotland after the Reformation; they were not specific to the western areas.

35. Blair, *Autobiography*, pp. 64, 84.

36. Livingstone, *Brief Historical Relation*, pp. 78–79.

37. Livingstone, *Brief Historical Relation*, p. 104.

38. Synod of Glasgow and Ayr, Minutes, April 1697, Scottish Record Office, CH2/464/1.

39. Session of St. John's, Perth, Minutes, Scottish Record Office, CH2/521/7–10; for an Irish example, see Session of Templepatrick, Minutes, 1646–1743, Public Record Office of Northern Ireland, CR4/12B/1.

40. Archbishop Bramhall's letter to William Laud, December 1634, as cited in Barkley, *Presbyterian Church in Ireland*, p. 3.

41. See Blair, *Autobiography*, p. 71; Livingstone, *Brief Historical Relation*, p. 79.

42. Livingstone, *Brief Historical Relation*, p. 80.

43. Stewart, "History of the Church of Ireland," pp. 320–21.

44. Livingstone, *Brief Historical Relation*, p. 78. In Carluke, Scotland, during the 1640s and 1650s, the same was required. See, for example, Carluke Kirk Session, Minutes, 27 June 1645, Scottish Record Office, CH2/56/1.

45. Sermon given by Thomas Orr while serving at a table during a Belfast communion. From a collection of sermon notes of John McBride and others, compiled by Daniel M.——, 1707, Presbyterian Historical Society, Belfast, located at Union Theological College, Box 1. See also Gabriel Sempil, Autobiography, MS copy (1728), National Library of Scotland MS 5746, pp. 22–23, for the discussion of Semple's communions during the 1650s.

46. Edmund S. Morgan: *Visible Saints, The History of a Puritan Idea* (Ithaca: Cornell Univ. Press, 1963) argues that during the 1620s and 1630s Puritans were just beginning to set evidence of conversion as a qualification for church membership. It does seem evident, however, that by 1636 Puritans in Massachusetts Bay had adopted a visible-saint church structure. My suggestion linking the Puritans' emphasis upon individual conversion to their rejection of the parish institutional structure is, and can only be, a hypothesis growing out of the Scots-Irish experience. The primary purpose of this discussion, however, is not an analysis of the Puritan religious community. Rather, I want to lay out the parallel conversion models of community and individual that characterized the Scots-Irish versus English Calvinist churches.

47. Livingstone, *Brief Historical Relation*, p. 67; see also Blair, *Autobiography*, p. 7.

48. John Thompson, "Westminster Confession," in *Crisis and Conflict, Essays in Irish Presbyterian History and Doctrine*, ed. J. L. M. Haire (Antrim, No. Ireland: W. & G. Baird, 1981), pp. 3–27.

49. Barkley, *Presbyterian Church in Ireland*, p. 8.

50. Robert Baillie, "Story of the Assemblie of Glasgow, to Spange" (1638), in *The Letters and Journals of Robert Baillie*, ed. David Laing (Edinburgh, 1841), 1:118–76.

51. A complete discussion of this rebellion is clearly beyond the scope of this chapter. Every history of Ireland discusses the 1641 rebellion at some length, but the fury of the period and the bias of the contemporary sources make a true rendition of the facts impossible. Yet even the most partisan accounts, emphasizing the contributing circumstances and the later atrocities of Cromwell's army, do not deny that, for example, a Protestant minister was actually crucified, or that women and even children were tortured and killed.

52. Adair, *True Narrative*, pp. 95–101.

53. See, for example, Associate Presbytery, *Act, Declaration, and Testimony*.

54. Robert Baillie, Correspondence with William Spange during the 1650s, in *The Letters and Journals of Robert Baillie*, vol. 3.

55. Baillie, "Story of the Assemblie," pp. 173–74.

56. Baillie to Spange, September 1640, in *The Letters and Journals of Robert Baillie*, 1:248–55.

57. Baillie to Spange, 20 August 1641, in *The Letters and Journals of Robert Baillie*, 1:360.

58. Baillie to Spange, 20 August 1641, in *The Letters and Journals of Robert Baillie*, 1:362.

59. Baillie to Spange, 20 August 1641, in *The Letters and Journals of Robert Baillie*, 1:358–77.

60. Baillie to Spange on the Assembly of 1642, in *The Letters and Journals of Robert Baillie*, 2:48.

Chapter 2

1. *Solemn League and Covenant* (Edinburgh, 1644).

2. John M. Barkley, *A Short History of the Presbyterian Church in Ireland* (Belfast: Presbyterian Church in Ireland, 1959).

3. *Minutes of the Sessions of the Westminster Assembly of Divines*, ed. Alex. F. Mitchell and John Struthers (Edinburgh: W. Blackwood & Sons, 1874), p. lxxx.

4. Antrim Meeting, Minutes, 1655, handwritten transcript by W. T. Latimer, 1900–1901, examined and corrected by John S. MacMaster, 1935, Public Record Office of Northern Ireland, D/1759/1A/1.

5. Antrim Meeting, Minutes, 1654–58; Templepatrick Session, Minutes, 1652–54, Public Record Office of Northern Ireland, CR4/12B/1. The stool of repentance was mentioned in the entry for 24 August 1647.

6. Antrim Meeting, Minutes, 5 November 1657.

7. Antrim Meeting, Minutes, 2 August 1654.

8. Robert Baillie to William Spange, 19 July 1654, in *The Letters and Journals of Robert Baillie*, ed. David Laing (Edinburgh, 1841), 3:244.

9. Baillie to Spange, 19 July 1654, in *The Letters and Journals of Robert Baillie*, 3:258–59.

10. Biographical note attached to Gabriel Sempil, Autobiography, MS copy (1728), National Library of Scotland MS 5746, pp. 22–23.

11. James Sharp to Duke of Lauderdale, 12 October 1660. All correspondence with the Duke of Lauderdale is from Lauderdale Papers, National Library of Scotland MS 2512.

12. Alexander Burnet to Lauderdale, 23 February 1664; Presbytery of Paisley, Minutes, 19 January 1664 and 28 January 1664, Scottish Record Office, CR2/294/5.

13. Sempil, Autobiography, pp. 28–54, citation, p. 53.

14. Presbytery of Lanark, Minutes, 1669–80, Scottish Record Office, CH2/234/2; Presbytery of Paisley, Minutes, 30 December 1666, 28 February 1667; Presbytery of Auchterarder, Minutes, 1 July 1674, Scottish Record Office, CH2/619/1.

15. Presbytery of Auchterarder, Minutes, 24 November 1682.

16. Presbytery of Auchterarder, Minutes, 3 October 1678.

17. Presbytery of Auchterarder, Minutes, 12 November 1678, 26 February 1679, 21 May 1679.

18. Presbytery of Auchterarder, Minutes, 14 September 1679.

19. Presbytery of Auchterarder, Minutes, 12 May 1680–8 December 1680.

20. Sharp to Lauderdale, 28 November 1662.

21. Burnet to Lauderdale, 23 February 1664.

22. Burnet to Lauderdale, 8 March 1664, 10 March 1664, 2 May 1664, 16 June 1664; citations from letters dated 21 January 1665 and 23 October 1665. In the letter dated 23 October 1665, Cruikshanks, no first name, is identified as the recent translater of Buchanan's *De jure regni apud Scotos*. However, I can find no bibliographic reference to this translation.

23. Sharp to Lauderdale, n.d. [summer or fall 1666]; Burnet to Lauderdale, 16 July 1667.

24. Sharp to Lauderdale, 19 May 1668.

25. Sharp to Lauderdale, 13 May 1674.

26. Burnet to Duke of Hamilton, 27 January 1676, Lauderdale Papers; citation, Burnet to Lauderdale, 15 August 1676.

27. Burnet to Lauderdale, 15 August 1676, 25 September 1679, 20 December 1679.

28. Burnet to Lauderdale, 22 January 1676.

29. George McKenzie to Col. James Douglas, 12 January 1685, 28 January 1685, 22 March 1685; William Patterson to Col. Douglas, 24 March 1685; citation, Lord Stormont to Douglas, n.d. [1685], National Library of Scotland MS 214.

30. Burnet to Lauderdale, 15 August 1676. See also Burnet to Lauderdale, 2 December 1665.

31. Burnet to Lauderdale, 23 October 1665.

32. Sharp to Lauderdale, 19 May 1668.

33. Sharp to Lauderdale, 13 May 1674.

34. Burnet to Lauderdale, 2 December 1665.

35. Burnet to Lauderdale, 15 August 1676.

36. Patrick Adair, *A True Narrative . . . of the Presbyterian Church in Ireland 1623–1670*, with introduction and notes by W. D. Killen (Belfast: C. Aitchison, 1866), pp. 258ff.

37. Ezekiel 37:10, 37:7–8; cited from King James Version.

38. Michael Bruce, *The Rattling of the Dry Bones, preached at Carluke May 1672* (n.p., [1672]), citations, pp. 3, 5.

39. Adair, *True Narrative*, pp. 258–61.

40. These laws include the Corporation Act of 1661, the Act of Uniformity of 1662, the Conventicle Act, 1664, and the Five Mile Act, 1665.

41. James Butler, Duke of Ormonde, as cited in James Seaton Reid, *History of the Presbyterian Church in Ireland* (London: Whittaker & Co., 1853), 2:284.

42. Laggan Meeting, Minutes, 1 March 1676, 24 April 1678, 24 April 1679, 9 July 1679, Presbyterian Historical Society, Belfast, located at Union Theological College, on shelf; Barkley, *Presbyterian Church in Ireland*, p. 48.

43. Laggan Meeting, Minutes, 21 August 1672.

44. Lord Chancellor Boyle to Ormonde, Dublin, 5 May 1667, Ormonde MSS, Public Record Office of Northern Ireland, vol. 3, fol. 26.

45. Thomas Otway to the Earl of Essex, 22 January 1676/7, Ormonde MSS, vol. 4, fol. 17. "Waistcoateers" were low-class prostitutes.

46. Laggan Meeting, Minutes, 16 May 1677, 8 August 1677.

47. Session of Burt, Minutes, 9 October 1678, 12 October 1678, 11 October 1679, 11 July 1680, 27 July 1681, 30 July 1681, 13 August 1682, Presbyterian Historical Society, Belfast, located at Union Theological College, Belfast, on shelf.

48. Unidentified official, cited in A. J. Weir, *Letterkenny—Congregation, Ministers and People 1615–1960* (photocopied booklet, n. d.).

49. Session of Burt, Minutes, 15 July 1694, 11 August 1696; see also 15 July 1697.

50. Laggan Meeting, Minutes, 14 July 1697; Presbytery of Route, Minutes, 3 July 1705, Presbyterian Historical Society, Belfast, located at Union Theological College, Box 1; Templepatrick Session, Minutes, 15 August 1697; Presbytery of Monaghan, Minutes, 14 July 1702, 30 November 1703, Presbyterian Historical Society, Belfast, located at Union Theological College, Box 10.

51. Session of Larne, Minutes, 15 July 1701, Presbyterian Historical Society, Belfast, located at Union Theological College, Box 9.

52. General Synod of Ulster, Minutes, 2 June 1697, printed in *Records of the General Synod of Ulster 1691–1820*, (Belfast: John Reid & Co., 1898), 1:22.

53. General Synod of Ulster, Minutes, 2 June 1697, 1:25.

54. Synod of Glasgow and Ayr, Minutes, 14 July 1692, Scottish Record Office, CH2/464/1.

55. Synod of Glasgow and Ayr, Minutes, 4 October 1692, [?] April 1697, CH2/464/1; 17 April 1709, 6 October 1709, CH2/464/2.

56. Presbytery of Paisley, Minutes, 20 November 1700, CH2/294/6.

57. Presbytery of Lanark, Minutes, 12 November 1702, CH2/234/4.

Chapter 3

1. Thomas Emlyn, *A True Narrative of the Proceedings of the Dissenting Ministers of Dublin against Mr. Thomas Emlyn* (London: J. Darby, 1719). See also Joseph Boyse, *A Vindication of the True Deity of our Blessed Saviour in Answer to a Pamphlet Entitled An Humble Inquiry into the Scripture-account of Jesus Christ, &c.* (Dublin, 1703), a response to a theological pamphlet by Emlyn.

2. Martin Tomkins, *The Case of Mr. Martin Tomkins* (London, 1719).

3. *The Anatomy of the Heretical Synod of Dissenters at Salters' Hall*, 2nd ed. (London, 1719), p. 2.

4. The first of the Thirty-Nine Articles of Religion reads as follows:
There is but one living and true God, everlasting, without body, parts, or passions; of infinite power, wisdom, and goodness; the Maker, and Preserver of all things both visible and invisible. And in unity of this Godhead there be three Persons, of one substance, power and eternity; the Father, the Son, and the Holy Ghost.

5. In addition to *Anatomy of the Heretical Synod*, see James Pierce's three pamphlets: *The Case of the Ministers Ejected at Exon*, *A Defence of the Case of the Ministers Ejected at Exon*, and *A Letter to Subscribing Ministers*, all published in London, 1719.

6. A. H. Drysdale, *History of the Presbyterians in England* (London: Publication Committee of the Presbyterian Church of England, 1889), p. 505.

7. This summary of positions, which fits the statements made by each in the pamphlets, comes from *An Impartial State of the Late Differences Amongst the Protestant Dissenting Ministers at Salters' Hall* (London, 1719).

8. *An Account of the Late Proceedings of the Dissenting Ministers at Salters' Hall in a Letter to the Rev^d Dr. Gale* (London, 1719), citation, p. 9.

9. *The Synod* (London, 1719), p. 5.

10. Thomas Bradbury, *An Answer to the Reproaches Cast on those Dissenting Ministers Who Subscrib'd their Belief of the Eternal Trinity* (London, 1719).

11. *An Account of Several Things* (London, 1719). Includes text of "Advices for Peace," pp. 1–11. For citations see pp. 25, 30–31. Also see *A True Relation of some Proceedings at Salters' Hall* (London [1719]).

12. Presbytery of Auchterarder, Minutes, 8 August 1717, Scottish Record Office, CH2/619/27. Included in this entry is a copy of the presbytery's paper to the General Assembly justifying its position.

13. Presbytery of Auchterarder, Minutes, 28 April 1713 to 28 January 1718.

14. *A Letter Concerning the Defections, Sins, and Backslidings of the Church of Scotland* (n.p., 1720).

15. The edition I have found useful is that with notes by Thomas Boston (Glasgow: John Bryce and David Patterson, 1722).

16. *A Letter to a Gentleman at Edinburgh, a Ruling Elder of the Church of*

Scotland, concerning the Proceedings of the last General Assembly, with reference unto Doctrine chiefly (n.p., n.d., text dated 20 April 1721).

17. The following books and pamphlets were published during the controversy: *A Letter to a Gentleman at Edinburgh*; *Queries, Agreed unto by the Commission of the General Assembly . . . Together with the Answers Given* (n.p., 1722); *Some Observations upon the Answers of the Brethren to the Queries* (Edinburgh, 1722); *Remarks upon the Answers of the Brethren to the Queries* (Edinburgh, 1722); *A Friendly Advice for Preserving the Purity of Doctrine and Peace of the Church* (Edinburgh: J. Mossman & Co., 1722); *Queries to the Friendly Advisor* (n.p., 1722); *The Protestation of Several Ministers of the Gospel, against the General Assembly's illegal Proceedings, upon the Head of Doctrine* (n.p., 1722); *Antinomianism of the Marrow of Modern Divinity Detected* (n.p., 1722); *A Letter to the Author of the Antinomianism of the Marrow of Modern Divinity detected* (Edinburgh, 1722).

18. *Advice for Preserving the Purity of Doctrine.*

19. *An Essay upon Gospel & Legal Preaching* (Edinburgh: James Davidson, 1723), pp. 4, 10, 14, 51, 68–69, 106.

20. [Ralph Erskine], *A Review of an Essay upon Gospel and Legal Preaching* (Edinburgh: Ja. McEuen, 1723).

21. Erskine, *Review of an Essay*, p. 5.

22. Laggan Presbytery, Minutes, 1672, Presbyterian Historical Society, Belfast, located at Union Theological College, on shelf.

23. General Synod of Ulster, Minutes, 3 June 1698, 5 June 1705, printed in *Records of the General Synod of Ulster, 1691–1820*, (Belfast: John Reid & Co., 1898), 1:34, 100.

24. General Synod of Ulster, Minutes, 19 June 1716, 1:396–97; citation, 1:396.

25. Francis Hutchison, as quoted in a letter from William Wright to Robert Wodrow, 27 September 1718. All correspondence with Robert Wodrow is located in Robert Wodrow Correspondence: Irish Letters 1705–1726, MS copies, Presbyterian Historical Society, Belfast, located at Union Theological College, Box 4.

26. John Abernethy, *Religious Obedience founded on Personal Persuasion* (Belfast: James Blow, 1720).

27. Abernethy, *Religious Obedience*, p. 9.

28. Abernethy, *Religious Obedience*, pp. 18, 20.

29. Abernethy, *Religious Obedience*, pp. 36, 37, 35–36.

30. General Synod of Ulster, Minutes, 21 June 1720, 1:521–22. For the perspective of the nonsubscribers, see *Narrative of the Proceedings of Seven General Synods . . . in which they issued a Synodical Breach* (Belfast: James Blow, 1727). This book not only gives the opposing view, for the synod record was made by the majority, but also includes magnificent detail on the actual substance of the debates.

31. General Synod of Ulster, Minutes, 21 June 1720, 1:535–37; citation, p. 536.

32. Robert McBride to James Stirling, 20 October 1720. All correspondence with James Stirling is located in Robert Wodrow Correspondence: Irish Letters.

33. Alexander McCracken to Stirling, 26 October 1720.

34. Joseph Boyse to Stirling, 13 December 1720.

35. Robert Craighead, *A Plea for Peace, or the Nature, Causes, Mischief, and Remedy of Church Divisions* (Dublin: George Grierson, 1721).

36. McCracken to Stirling, n.d., [between December 1720 and April 1721].

37. George Lang to Wodrow, 23 May 1721.

38. General Synod of Ulster, Minutes, 23 June 1721, 2:8–9.

39. General Synod of Ulster, Minutes, 23 June, 24 June, 27 June 1721, 2:9–11, 13–14.

40. General Synod of Ulster, Minutes, 24 June 1721, 2:11.

41. General Synod of Ulster, Minutes, 27 June 1721, 2:14–15.

42. General Synod of Ulster, Minutes, 28 June 1721, 2:16–17.

43. James Seaton Reid, *History of the Presbyterian Church in Ireland*, 2nd ed. (London: Whittaker & Co., 1853). Reid refers to more than sixty pamphlets that he has read in his research of this era. I have found about thirty and have read the following: *A Vindication of the Presbyterian Ministers in the North of Ireland; Subscribers and Nonsubscribers. Published by Victor Ferguson* (Belfast: James Blow, 1721); [John Abernethy], *Seasonable Advice to the Protestant Dissenters in the North of Ireland*, with preface by Nath. Weld, J. Boyse and R. Choppin (Dublin: James Carson, 1722); [Samuel Hemphill], *Some General Remarks Argumentative and Historical on the Vindication Publish'd by Dr. Ferguson* (n.p., 1722); Matthew Clerk, *A Letter from the Belfast Society to the Rev. Matthew Clerk . . . with an Answer to the Society's Remarks* ([Belfast] 1723); John Abernethy, *A Defence of the Seasonable Advice* (Belfast: James Blow, 1724); Samuel Haliday, *Reasons Against the Imposition of Subscription to the Westminster Confession of Faith* (Belfast: James Blow, 1724); Gilbert Kennedy, *A Defence of the Principles and Conduct of the Rev'd General Synod of Ulster* (Belfast: Robert Gardner, 1724); James Kirkpatrick, *A Scriptural Plea Against a Fatal Rupture and Breach of Christian Communion* (Belfast, 1724); Thomas Nevin, *The Tryal of Thomas Nevin* (Belfast: James Blow, 1725); C. Masterton, *Christian Liberty founded in Gospel Truth* (Belfast: Robert Gardner, 1725); Samuel Haliday, *A Letter to the Reverend Mr. Gilbert Kennedy* (Belfast: James Blow, 1725); Robert Higginbotham, *Reasons Against the Overtures* (Belfast: James Blow, 1726); Robert McBride, *The Overtures transmitted by the General Synod, 1725, Set in a Fair Light* (Belfast: Robert Gardner, 1726).

44. *A Vindication by Victor Ferguson*, pp. 14, 42–43, 53.

45. Masterton, *Christian Liberty in Gospel Truth*, pp. 14, 15, 32.

46. Abernethy, *Seasonable Advice*, pp. 5, 9.

47. Lang to Wodrow, 4 November 1721.

48. See [Samuel Hemphill], *Some General Remarks on the Vindication*. See also Clerk, *A Letter from the Belfast Society with an Answer.*

49. General Synod of Ulster, Minutes, 20 June 1723, 2:45.

50. General Synod of Ulster, Minutes, June 1723, 2:50–65. For a detailed account from the other side, see *Proceedings of Seven General Synods*, pp. 93–133.

51. Charles Masterton to Wodrow, 12 July 1723. See also from the Wodrow correspondence Masterton to Wodrow, 12 November 1723 and Masterton to William Macknight, 15 January 1724.

52. William Livingston to William Macknight, 24 April 1724, in Robert Wodrow Correspondence: Irish Letters.

53. General Synod of Ulster, Minutes, June 1724, 2:69–73, 76–81; citation, p. 80. See also Nevin, *The Tryal of Thomas Nevin*.

54. Andrew Gray to Wodrow, 6 May 1725. See the same writer to same correspondent 29 December 1724 and 22 February 1725 for the entire Colville story from the English perspective.

55. General Synod of Ulster, Minutes, 23 June 1725, 2:88–89, 91, 94, 18, 22.

56. Masterton to Wodrow, 1 November 1725.

57. General Synod of Ulster, Minutes, June 1725, 2:97.

58. General Synod of Ulster, Minutes, June 1726, 2:103–5, 108–9; *Proceedings of Seven General Synods*, pp. 180–299; Masterton to William Macknight, 1 July 1725, in Robert Wodrow correspondence: Irish Letters; Antrim Presbytery, *A Letter from the Presbytery of Antrim to the Congregations under their Care: Occasion'd by the Uncharitable Breach* (Belfast: James Blow, 1726).

59. McBride, *Overtures Set in a Fair Light* and J. Boyse, *A Vindication of a Private Letter* (Belfast: James Blow, 1726).

60. An excellent summary of the cases presented by each side and the apparent focus of debate may be seen in A. W. Godfrey Brown, "A Theological Interpretation of the First subscription Controversy (1719–1728)" in *Challenge and Conflict, Essays in Irish Presbyterian History and Doctrine*, ed. J. L. M. Haire (Antrim, No. Ireland: W. & G. Baird, 1981), pp. 28–45.

61. *Proceedings of Seven General Synods*, pp. 192–93.

62. *Proceedings of Seven General Synods*, p. 92.

63. As cited in David Stewart, *The Seceders in Ireland* (Belfast: Presbyterian Historical Society in Ireland, 1950), p. 39.

Chapter 4

1. Session of Cambuslang, Minutes, 20 April 1720, Scottish Record Office, CH2/415/2.

2. Presbytery of Auchterarder, Minutes, 3 May 1723, 18 May 1723, 25 June 1723, Scottish Record Office, CH2/619/27.

3. Presbytery of Auchterarder, Minutes, 25 February 1729.

4. Presbytery of Hamilton, Minutes, 25 February 1729, Scottish Record Office, CH2/393/3.

5. Presbytery of Lanark, Minutes, 12 March 1729, Scottish Record Office, CH2/234/7.

6. Presbytery of Paisley, Minutes, 28 January 1729, 26 February 1729, 18 March 1729, Scottish Record Office, CH2/294/8.

7. Reference from the Presbytery of St. Andrews to the General Assembly, signed by Fran: Gray, clerk, 26 October 1726, Lee Papers, vol. 2, fol. 5, National Library of Scotland MS 3431.

8. As cited in Presbytery of Auchterarder, Minutes, 13 May 1731.

9. Presbytery of Auchterarder, Minutes, 13 May 1731.

10. In Scotland the issue of patronage was further complicated by the fact that heritors or sessions frequently acted in behalf of the congregation. The role of the congregation as distinct from its representatives was not clearly established, nor were basic lines of authority. Sometimes the people challenged the sessions, sessions challenged heritors, or several heritors challenged their more prestigious colleagues. If the patron became involved, those challenging the patron's candidate would complain of the patron's imposition. For examples of pamphlets on either side of this issue see [David Dalrymple], *An Account of Law Patronage in Scotland* (Edinburgh: J. M., 1712) against patronage and *Remarks upon the Representation Made by the Kirk of Scotland Concerning Patronage* (n.p., n.d. [probably 1712]) in favor.

11. *Act of the General Assembly of the Church of Scotland concerning the Method of Planting Vacant Congregations* (Edinburgh: J. Davidson & R. Fleming, 1732).

12. Presbytery of Paisley, Minutes, 12 July 1731, 1 September 1731, 22 September 1731, 5 October 1731, 28 October 1731, 16 February 1732, 6 April 1732, 26 April 1732.

13. Presbytery of Auchterarder, Minutes, 4 April 1732.

14. *A Public Testimony* (Edinburgh: Thomas Lumisden and John Robertson, 1732), citation, p. 11.

15. For details of the debate, see: *The State and Duty of the Church of Scotland, especially with Respect to the Settlement of Ministers* (Edinburgh: W. Cheyne, 1732); *The Overture anent Planting vacant Parishes* (Edinburgh, 1732); *An Enquiry into the Methods of Settling Parishes* (Edinburgh, 1732); *The Overture considered* (Edinburgh: Thomas Lumisden and John Robertson, 1732); *A seasonable Warning to all Lovers of Jesus Christ, Members of the Church of Scotland* (Glasgow, 1733); *The Defection of the Church of Scotland from her Reformation-Principles Considered* (Edinburgh: Thomas Lumisden and John Robertson, 1733).

16. Associate Presbytery, Minutes, 6 December 1733, MS copy certified by James Hall, 1824, Scottish Record Office, CH3/27/1.

17. *Act of the General Assembly of the Church of Scotland, concerning the Ministers seceding from the said Church* (n.p., 1735). See also *Copy of a Libel agst. Messrs. Ebenezer Erskine and Others, Ministers, who have seceded from the Church of Scotland (n.p., 1739).*

18. *Act of the General Assembly concerning the Ministers seceding.*

19. Associate Presbytery, Minutes, 17 December 1735, 5 January 1737, 15 May 1739, 13 June 1739, 27 December 1739, 22 July 1740.

20. Presbytery of Auchterarder, Minutes, 30 May 1732, 6 March 1733, 30 March 1733, 21 August 1733, 11 December 1733.

21. Presbytery of Auchterarder, Minutes, 19 March 1734, 10 April 1734, 16 July 1734, 7 August 1734, 24 December 1734, 21 January 1735, 18 November 1735, 8 February 1737, 1 November 1737, 14 March 1738, 10 June 1739.

22. Hew Scott, *Fasti Ecclesiae Scoticanae* (Edinburgh: Tweeddale Court, 1925), 5:69.

23. Session of St. John's, Perth, Minutes, 21 July 1733, 14 December 1738, Scottish Record Office, CH2/521/17.

24. Associate Presbytery, Minutes, 5 November 1736.

25. Associate Presbytery, *Act, Declaration, and Testimony for the Doctrine, Worship, Discipline, and Government of the Church of Scotland* (Edinburgh: Thomas Lumisden and John Robertson, 1737).

26. Associate Presbytery, Minutes, 9 January 1741.

27. Associate Presbytery, Minutes, 9 January 1741.

28. David Scott, *Annals and Statistics of the Original Secession Church* (Edinburgh: Andrew Elliot, 1886), pp. 19–20.

29. Associate Presbytery, Minutes, 28 July 1736.

30. Scott, *Annals of the Secession Church*, pp. 11–12.

31. Scott, *Annals of the Secession Church*, p. 20.

32. Associate Presbytery, Minutes, 22 December 1737.

33. Scott, *Annals of the Secession Church*, pp. 16–17.

34. Associate Presbytery, Minutes, 4 September 1736.

35. David Stewart, *The Seceders in Ireland* (Belfast: Presbyterian Historical Society in Ireland, 1950), p. 231.

36. General Synod of Ulster, Minutes, printed in *Records of the General Synod of Ulster*, (Belfast: John Reid & Co., 1898), June 1717, 1:427–49; June 1719, 1:458–64, 471; June 1719, 1:488–94, 506; June 1720, 1:523–33.

37. General Synod of Ulster, Minutes, June 1727, 2:117.

38. Presbytery of Killyleagh, Minutes, 27 September 1727–16 January 1728, Presbyterian Historical Society, Belfast, at Union Theological College, Box 3.

39. General Synod of Ulster, Minutes, June 1727, 2:117, 119.

40. Stewart, *Seceders in Ireland*, p. 256.

41. Stewart, *Seceders in Ireland*, pp. 251–52; General Synod of Ulster, Minutes, June 1743, 2:490–91.

42. Stewart, *Seceders in Ireland*, pp. 265–66.

43. General Synod of Ulster, Minutes, June 1733, 2:185; June 1737, 2:194–95; June 1735, 2:206–7; June 1736, 2:216.

44. See Session of Lisburn, Minutes, 1717–1720 for excellent examples of creative quarrelsomeness between elders and their pastor. Presbyterian Historical Society, Belfast, located at Union Theological College, Box 11.

45. Session of Lisburn, Minutes, 1731–36; General Synod of Ulster, Minutes, June 1731, 2:172; June 1732, 2:175–76; June 1733, 2:185–86; June 1734, 2:191; June 1735, 2:204–6; June 1736, 2:217; June 1737, 2:228.

46. Stewart, *Seceders in Ireland*, pp. 261–62.

47. James Seaton Reid, *History of the Presbyterian Church in Ireland*, 2nd ed. (London: Whittaker & Co., 1853), vol. 2.

48. Stewart, *Seceders in Ireland*, pp. 261–62; General Synod of Ulster, Minutes, June 1744, 2:305–6; June 1745, 2:312–13.

49. Included in General Synod of Ulster, Minutes, June 1747, 2:330–31.

50. Presbytery of Strabane, Minutes, 19 November 1729–7 November 1739, Presbyterian Historical Society, Belfast, located at Union Theological College, on shelf; General Synod of Ulster, Minutes, June 1730, 2:153; June 1732, 2:177; June 1733, 2:182–83; June 1740, 2:258–60; June 1741, 2:266; June 1734, 2:192; June 1735, 2:207–8; June 1736, 2:215–16.

51. Associate Presbytery, Minutes, 16 June 1742.

52. For a general account of the revival sparked by Whitefield's tour of Scotland there are two reasonable secondary accounts. D. MacFarlan, *The Revivals of the Eighteenth Century particularly at Cambuslang* (1847; rpt., Wheaton, Ill.: Richard Owen Roberts, 1980) provides a decent narrative, though his selection and use of sources should be approached with caution. Far more reliable is Arthur Fawcett, *Cambuslang Revival* (Edinburgh: Banner of Truth Trust, 1971). Although Fawcett's interpretation is clearly biased in favor of the character, strength, and success of the revivals, his fidelity to the rules of historical evidence render Fawcett a good source. Fawcett argues, and I agree, that the coming of Whitefield and the outbreak of revivalism inhibited the expansion of the Secession Church.

53. James Robe, *A Faithful Narrative . . . of Kilsyth* (Edinburgh, 1742), citation, p. 25.

54. *A Short Narrative . . . Cambuslang* (Glasgow: William Duncan, 1742). Extensive records of the revival at Cambuslang were kept by the congregation's pastor, William MacCulloch. These are the primary sources used by Arthur Fawcett in his account.

55. Alexander Webster, *Divine Influence . . . at Cambuslang* (Edinburgh: Thomas Lumisden and John Robertson, 1742), p. 4.

56. Robe, *Faithful Narrative*, p. 45.

57. *Short Account . . . of Cambuslang* (Glasgow: Robt. Smith, 1742), citations pp. 3, 4, 6, 13.

58. James Robe, *Mr. Robe's first Letter to the Rev.[d] James Fisher* (Glasgow, 1742); Robe, *Second Letter . . .* (Edinburgh, 1742); Robe, *Third Letter . . .* (Edinburgh, 1743); Robe, *Fourth Letter . . .* (Edinburgh, 1743). Citations are from the *Third Letter*, pp. 3–4, 48.

59. *A Letter from Mr. John Willison . . . to Mr. James Fisher* (Edinburgh: Thomas Lumisden and John Robertson, 1743).

60. *Observations in Defense of the Work at Cambuslang* (Edinburgh, 1742).

61. Robe, *Faithful Narrative*.

62. Session of Cambuslang, Minutes, 18 July 1742.

63. Robe, *Faithful Narrative*, pp. 99–108.

Chapter 5

1. Richard Webster, *A History of the Presbyterian Church in America, From Its Origin Until the Year of 1760* (Philadelphia: Joseph M. Wilson, 1857), pp. 76–78; Leonard J. Trinterud, *The Forming of An American Tradition* (Philadelphia: Westminster Press, 1949), pp. 22–26. The best general history of the Presbyterian church in the eighteenth century is Trinterud's work; he has done an excellent job of piecing together a smooth narrative from the bits of original manuscripts and the early nineteenth-century histories. He is especially dependent upon Richard Webster's work, whose excellent biographical sketches of the eighteenth-century ministers are an impressive compilation from numerous and diverse sources. Webster brings a fine, critical eye to his evidence, thus his history continues to stand as an excellent source book for modern scholars.

2. Letter from Bishop of Kilmore and Armagh to [?], dated London, 6 January 1712/13, based on reports from Dean of Kilmore and Archdeacon Handcock. Crossle MSS Papers, Carter I, fol. 56. See also [no first name] Phipps to the Chancellor, Dublin, 23 December 1712, vol. 11, Country Letters; and letter from Francis Iredell to Dean of Kilmore, 24 August 1713, MS copy T808. All collections listed are held by the Public Record Office of Northern Ireland.

3. Letter from Francis Iredell to Robert Craghead, January 1728/29, MS copy, Public Record Office of Northern Ireland, T808.

4. Report of John St. Leger and Michael Ward to the Lord Justices, 26 June 1729, Public Record Office of Northern Ireland, D2092/1/3.

5. Ezekial Stewart to Michael Ward, 25 March 1729, Public Record Office of Northern Ireland, D2092/1/3.

6. Letter by James Murray printed in *Pennsylvania Gazette*, 27 October–3 November 1737. Citation from the 3 November segment.

7. There remains much research to be done on the migration of Ulster Scots to the colonies during the years 1700–1775. An excellent, interpretive summary of scholarship to date can be found in Kerby A. Miller, "Settlers, Servants, and Slaves: Irish Emigration before the American Revolution," in *Emigrants and Exiles: Ireland and the Irish Exodus to North America* (New York: Oxford Univ. Press, 1985), pp. 137–68. See also R. J. Dickson, *Ulster Emigration to Colonial America, 1718–1775* (London: Routledge & Kegan Paul, 1966).

8. Thomas Whitney to [?], 27 July 1728, Public Record Office of Northern Ireland, T808; Lord Carteret to the Lords Justices of Ireland, 2 November 1728, Public Record Office of Northern Ireland, T808; Boulter and Gambie as cited in Guy Soulliard Klett, *Presbyterians in Colonial Pennsylvania* (Philadelphia: Univ. of Pennsylvania Press, 1937), pp. 19, 20.

9. *Pennsylvania Gazette*, 9 September 1736; Arthur Young, *A Tour of Ireland . . . made in the years 1776, 1777 and 1778*, as cited in Klett, *Presbyterians in Pennsylvania*, p. 22; Miller, *Emigrants and Exiles*, pp. 153–55; Samuel Blair, as cited in Webster, *Presbyterian Church to 1760*, p. 120.

10. Francis Makemie, *Life and Writings of Francis Makemie*, ed. Boyd

Schlenther, (Philadelphia: Presbyterian Historical Society, 1975), pp. 1–28. Francis Makemie has become a legendary figure in the American Presbyterian tradition as "Founder of the Church"; thus every history includes some discussion of his work. Schlenther's biographical essay is the most extensive, recent treatment of Makemie that I have found. It includes a thorough summary of what others have said, as well as analyses of Makemie's own writings.

11. Biographical data on ministers comes from three sources. Hew Scott, *Fasti Ecclesiae Scoticanae* (Edinburgh: Tweeddale Court, 1927), vols. 1–7, and James McConnell, *Fasti of the Irish Presbyterian Church 1613–1840*, revised by Samuel G. McConnell (Belfast: Presbyterian Historical Society, 1951) are biographical indices of all ministers ordained in Scotland and Ireland, respectively. They are, in effect, a comprehensive collection of all information regarding each minister who can be found in official assembly, synod, and presbytery records; these sketches are often fleshed out from other sources as well. The third work I have used is Webster, *Presbyterian Church to 1760*, which includes biographical sketches of all colonial Presbyterian ministers, again gleaned from official sources.

12. Francis Makemie to Benjamin Coleman, 28 March 1707, in *Writings of Francis Makemie*, pp. 252–53. Neither the sermons nor the comments were ever recorded in the documents. These sermons, based upon sequential verses of the Letter to the Hebrews, continued until 1717, when the presbytery first met as only one of four under the jurisdiction of the Synod of Philadelphia.

13. Webster, *Presbyterian Church to 1760*, p. 311. A fourth presbytery, Snow Hill, formed to include the three Maryland clerics, was never established due to the death in 1718 of John Henry of Rehoboth and the ill health of John Hampton.

14. Presbytery of Philadelphia, Minutes, 1708–1712, printed in *Minutes of the Presbyterian Church in America*, ed. Guy S. Klett, (Philadelphia: Presbyterian Historical Society, 1976), pp. 5, 7, 8, 9, 12, 13, 15–16. Also printed in this collection are two letters from the Presbytery of Philadelphia to Woodbridge, September 1710, pp. 74–75; Presbytery of Philadelphia to Woodbridge, September 1711, pp. 78–79; Presbytery of Philadelphia to Cotton Mather, September 1712, pp. 79–81; Presbytery of Philadelphia to Nathanial Wade, September 1712, p. 81; Presbytery of Philadelphia to Woodbridge, September 1712, p. 82.

15. Joseph Morgan, *The Great Concernment of Gospel Ordinances Manifested from the great effect of the well Improving or Neglect of Them* (New York: William and Andrew Bradford, 1712), pp. 12, 9.

16. Synod of Philadelphia, Minutes, 27 September 1721, printed in *Minutes of the Presbyterian Church*, p. 51.

17. Synod of Philadelphia, Minutes, 30 September 1720, pp. 46–47; 25 September 1721, p. 50.

18. Jedidiah Andrews to Benjamin Coleman, 30 April 1722, copy at Presbyterian Historical Society, Philadelphia, original at American Antiquarian Society.

19. Jonathan Dickinson, *A Sermon preached At the Opening of the Synod at Philadelphia September 19, 1722* (Boston: T. Fleet for S. Gerrish, 1728), citations, pp. 3, 12, 17.

20. Dickinson, *Sermon . . . September 19, 1722*, p. 18.

21. Synod of Philadelphia, Minutes, 27 September 1722, pp. 57–58.

22. There is no record of such attempts in the synod minutes, but Jedidiah Andrews reported this in a letter to Benjamin Coleman, 7 April 1729, copy at Presbyterian Historical Society, Philadelphia, original at Massachusetts Historical Society.

23. Synod of Philadelphia, Minutes, 18 September 1724, pp. 64–65. After the 1729 meeting, the synod meeting was always a full synod.

24. Andrews to Coleman, 7 April 1729.

25. John Thompson, *An Overture Presented to the Rev. Synod of dissenting Ministers Sitting in Philadelphia, in the Month of September 1728* (Philadelphia, 1729), citation, pp. 30–31.

26. Thompson, *Overture, 1728*, pp. 7–8, 23–24.

27. Thompson, *Overture, 1728*, pp. 12, 4.

28. Jonathan Dickinson, *Remarks Upon a Discourse intituled An Overture Presented to the Reverend Synod of Dissenting Ministers sitting in Philadelphia, in the Month of September 1728* (New York: J. Peter Zenger, 1729), pp. 13–15, citation p. 14.

29. Dickinson was an active spokesman for the Presbyterian way against other denominational options. See, for example, his *A Defence of Presbyterian Ordination* (Boston: for Daniel Henchman, 1724), *The Scripture Bishop Or the Divine Right of Presbyterian Ordination and Government* (Boston: for Daniel Henchman, 1732), *The Reasonableness of Non-Conformity to the Church of England* (Boston: Kneeland and Green, 1738), and *Brief Illustration and Confirmation of the Divine Right of Infant Baptism* (Boston: S. Kneeland & T. Green, 1746), this last against the Baptists.

30. Dickinson, *Remarks Upon a Discourse*, p. 20.

31. Dickinson, *Remarks Upon a Discourse*, pp. 4, 8, 20, 23–25, 9.

32. Dickinson, *Remarks Upon a Discourse*, pp. 4–6.

33. Synod of Philadelphia, Minutes, September 1729, pp. 103–4.

34. Webster, *Presbyterian Church to 1760*, pp. 410–11. The *Fasti* recorded no mention of Orr's involvement with the Ulster Synod. This apparent contradiction may indicate that Orr never progressed far enough in his studies in Ireland to involve the synod. Had a presbytery found him unacceptable, this would certainly have been noted. Usually not until a minister was licensed was he recorded in synod minutes, and Irish presbytery records are too scattered for a complete inventory. Of course, considering the final outcome of the Orr dispute, students like Orr may have been the subject of colonial synodical concerns regarding the acceptance of unquestioned students from Ireland.

35. Presbytery of Donegal, Minutes, 2, 4 April 1734, Presbyterian Historical Society, Philadelphia; Synod of Philadelphia, Report of Commission Meeting 7 November 1734, in Minutes, 18 September 1735, pp. 125–27.

36. Presbytery of Donegal, Minutes, 11–12 June 1735.

37. Presbytery of Donegal, Minutes, 2–5 September 1735.

38. When a minister left the bounds of his presbytery, hoping to serve as a minister within another jurisdiction of the same communion, he carried a certificate from his presbytery of previous residence. The certificate declared him a cleric of orthodox beliefs and upstanding moral conduct. Without a certificate, new presbyteries refused membership. Thus in this action the presbytery essentially prohibited Orr from preaching or serving anywhere in the Presbyterian community, including in Britain, unless he chose to begin again in a place where he was not known.

39. Presbytery of Donegal, Minutes, 8–10 October 1735, 9 November 1735, 14 April 1736, 13 October 1736.

40. Sub-Synod of Londonderry, Minutes, 13 May 1735, Presbyterian Historical Society, Belfast, located at Union Theological College, Box 4.

41. General Synod of Ulster, Minutes, June 1735, printed in *Records of the General Synod of Ulster, 1691–1820* (Belfast: John Reid & Co., 1898), 2:308.

42. Ebenezer Pemberton, *A Sermon Preached Before the Commission of the Synod* (New York: John Peter Zenger, 1735), citations, pp. 3, 15–16.

43. George Gillespie, *Treatise Against the Deists, or Free Thinkers* (Philadelphia: A. Bradford, 1735).

44. As cited in Benjamin Franklin, *Some Observations on the Proceedings Against the Rev. Mr. Hemphill* (Philadelphia: B. Franklin, 1735), p. 17.

45. Franklin, *Observations on Hemphill,* p. 13. See also, Benjamin Franklin, *A Defense of Mr. Hemphill's Observations* (Philadelphia: B. Franklin, 1735); Franklin, *A Letter to a Friend in the Country* (Philadelphia: B. Franklin, 1735).

46. George Whitefield, Journal, 13 November 1739, printed in *Journals* [1738–47] (Edinburgh: Banner of Truth Trust, 1960), p. 347.

47. Andrews to Coleman, 14 June 1735, as cited in Webster, *Presbyterian Church to 1760,* p. 417 (original correspondence at American Antiquarian Society).

48. Jonathan Dickinson, *Remarks Upon a Pamphlet Entitled A Letter to a Friend in the Country* (Philadelphia: Andrew Bradford [1735]).

49. Franklin later claimed that he supported Hemphill despite the plagiarism: "I rather approv'd his giving us good Sermons compos'd by others, than bad ones of his own Manufacture; tho' the latter was the Practice of our common Teachers." *Benjamin Franklin's Autobiography,* ed. J. A. Leo LeMay and P. M. Zall (New York: Norton, 1986), p. 82.

50. Synod of Philadelphia, Minutes, 22 September 1735, pp. 131–33.

51. Klett, *Presbyterians in Pennsylvania,* pp. 108 ff.; Trinterud, *Forming of A Tradition,* pp. 180–82.

52. *Sermons on Sacramental Occasions by Divers Ministers* (Boston: J. Draper for D. Henchman, 1739).

53. Gilbert Tennent, Sermon One, in *Sermons on Sacramental Occasions,* citations, pp. 21, 22.

54. Gilbert Tennent, Sermon Two, in *Sermons on Sacramental Occasions,* pp. 52–57, citation, p. 57. See also Gilbert Tennent, Sermon Five.

55. Samuel Blair, Sermon Three, in *Sermons on Sacramental Occasions.*

56. William Tennent, Sermon Four, in *Sermons on Sacramental Occasions*, citation, p. 108.

57. Presbytery of Donegal, Minutes, 26 May 1735–5 October 1737.

58. Franklin, *Defense of Hemphill's Observations*, p. 18.

Chapter 6

1. James McConnell, *Fasti of the Irish Presbyterian Church*, revised by Samuel G. McConnell (Belfast: Presbyterian Historical Society, 1951); Synod of Philadelphia, Minutes, 17 September 1718, printed in *Minutes of the Presbyterian Church in America, 1706–1789*, ed. Guy S. Klett (Philadelphia: Presbyterian Historical Society, 1976), p. 34.

2. The details of Gilbert Tennent's career can be found in both McConnell's *Fasti* and the minutes of the Synod of Philadelphia.

3. George Whitefield, Journal, 22 November 1739, printed in *Journals* [1738–47] (Edinburgh: Banner of Truth Trust, 1960), p. 354; 14 November 1739, p. 348.

4. Samuel Finley, *The successful Minister of Christ distinguished in Glory* ([Philadelphia: William Bradford, 1764]), pp. 19, 11.

5. Synod of Philadelphia, Minutes, 29 May 1738, p. 157.

6. Synod of Philadelphia, Minutes, 26 May 1738, p. 154.

7. Whitefield, Journal, 22 November 1739, p. 354.

8. Presbytery of New Brunswick, Minutes, 8 August 1738, Presbyterian Historical Society, Philadelphia.

9. John Rowland, *A Narrative of the Revival and Progress of Religion* (Philadelphia: William Bradford, 1745), p. 40. More telling is Tennent's follow-up comment that Rowland "excell'd in thos [Endowments] of a gracious Nature." These gracious qualifications became the New Light response to the criticism of the Log College.

10. Presbytery of New Brunswick, Minutes, 7 September 1738; Presbytery of Philadelphia, Minutes, 1 March 1737/38, 19 September 1738, 26 October 1738, Presbyterian Historical Society, Philadelphia; Synod of Philadelphia, Minutes, 29 May 1739, p. 164.

11. Synod of Philadelphia, Minutes, 29 May 1739, pp. 165–66; Minutes of the Commission of the Synod included in Synod of Philadelphia, Minutes, 15 August 1739, pp. 166–67.

12. My greatest argument with Leonard J. Trinterud's *The Forming of An American Tradition* (Philadelphia: Westminster Press, 1949) is precisely this. He perceived the Old Lights as morally lax and needlessly nasty, while in his view the New Lights only occasionally went overboard. The latter were justified because they were fighting a strong, unfair opponent. As will be seen, I believe that just as the Old Lights held the balance of institutional power, the New Lights had the attention of the laity. Both sides operated from positions of both weakness and strength, and each was struggling for survival.

13. Synod of Philadelphia, Minutes, 26 May 1738, p. 155.

14. Gilbert Tennent, *The Danger of An Unconverted Ministry* (Philadelphia: Benjamin Franklin, 1740).

15. Tennent, *Danger of Unconverted Ministry*, pp. 3–7.

16. Tennent, *Danger of Unconverted Ministry*, pp. 9–16, citations, pp. 10, 16.

17. Tennent, *Danger of Unconverted Ministry*, pp. 17–30.

18. David Evans, *The Minister of Christ And the Duties of his Flock* (Philadelphia: B. Franklin, 1732), pp. 59–60.

19. Synod of Philadelphia, Minutes, 17 September 1736, p. 142.

20. Jonathan Dickinson, *The Danger of Schisms and Contentions with Respect to the Ministry and Ordinances of the Gospel* (New York: J. Peter Zenger, 1739), citations pp. 8, 17.

21. Gilbert Tennent, *Die Gefahr bey unbekehrten Predigern* (Germantown: Christoph Saur, 1740).

22. Evans, *Minister of Christ*, pp. 7–17, citation, pp. 9–10.

23. Synod of Philadelphia, Minutes, 2 June 1740, p. 171. For two excellent presentations on this problem from the New Light vantage point, see Jonathan Dickinson, *The Nature and Necessity of Regeneration* (New York: James Parker, 1743) and Gilbert Tennent, *The Terrors of the Lord* (Philadelphia: William Bradford, 1749).

24. Gilbert Tennent to Thomas Prince, in *The Christian History* (Boston: S. Kneeland and T. Green for T. Prince, 1745), p. 293.

25. Trinterud, in *Forming of A Tradition*, was among the first to discuss the current issue of German influence on the Tennents. No one disputes Gilbert Tennent's correspondence with Frelinghuysen; the question is one of influence. As Tennent had not been a pastor long enough to either change or maintain a style of ministry, it is difficult to analyze the German cleric's effect. Frelinghuysen could easily have been a catalyst. On the other hand, it will be seen that certain patterns in Tennent's approach and style follow models established in Ulster and Scotland. From my perspective, this issue is less important than the question of lay response. The evangelical preachers were not, after all, successful everywhere.

26. John Tennent, *The Nature of Regeneration Opened* (Boston, 1735), citation, p. 2; details regarding the revival are found in the preface to these sermons written by Gilbert Tennent, pp. i-xv. For a full discussion of this first revival, see Ned Landsman, "Revivalism and Nativism in the Middle Colonies: The Great Awakening and the Scots Community in East New Jersey," *American Quarterly*, 34 (1982), 149–64.

27. See the following sermons by Gilbert Tennent: *The Danger of Forgetting God describ'd* (New York: John Peter Zenger, 1735); *The Espousals or a Passionate Perswasive* (New York: J. Peter Zenger, 1735); *The Necessity of Religious Violence in Order to Obtain Durable Happiness* (New York: William Bradford, 1735); *A Solemn Warning to the Secure World from the God of terrible Majesty* (Boston: S. Kneeland and T. Green for D. Henchman, 1735); *The Necessity of Receiving the Truth in Love* (New York: John Peter Zenger, 1735).

28. Finley, *The Successful Minister*, p. 23.

29. Tennent, *Solemn Warning*, p. viii.

30. Tennent, *Solemn Warning*, p. 10.

31. Tennent, *Solemn Warning*, p. 36.

32. Tennent, *Danger of Forgetting God*, p. 8; *The Espousals*, p. 22.

33. Synod of Philadelphia, Minutes, 30 May 1737, p. 150.

34. Synod of Philadelphia, Minutes, 26 May 1738, p. 153; 28 May 1739, p. 163.

35. Rowland, *Narrative of the Revival*; Gilbert Tennent, *A Funeral Sermon, Occasioned by the Death of the Reverend Mr. John Rowland*, (Philadelphia: William Bradford, 1745), citation, pp. 40–41. Boanerges was a surname given by Jesus to James and John as sons of thunder.

36. Dickinson, *Danger of Schisms*, pp. 6, 24, 25, 12.

37. Presbytery of Donegal, Minutes, 4 September 1740, 5 November 1740, Presbyterian Historical Society, Philadelphia.

38. Presbytery of Donegal, Minutes, 6 November 1740, 10–11 December 1740.

39. David Evans, *Law and Gospel Or Man wholly Ruined by the Law and Recovered by the Gospel* (Philadelphia: B. Franklin, 1748).

40. Whitefield, Journal, 6 November 1739, p. 342.

41. Whitefield, Journal, 8 November 1739, p. 343; 14 November 1739, p. 349; 21 November 1739, p. 353; 28 November 1739, p. 359.

42. Whitefield, Journal, 8 November 1739, p. 343.

43. Whitefield, Journal, 29 November 1739, pp. 361–62.

44. Benjamin Franklin, *Benjamin Franklin's Autobiography*, ed. J. A. Leo LeMay and P. M. Zall (New York: Norton, 1986), 87, 88.

45. Anyone interested in the extent of George Whitefield's middle-colony tours should read his Journal, 30 October–2 December 1739, pp. 338–64, and 13 April–14 May 1740, pp. 405–27.

46. William Howland Kenney, "George Whitefield and Colonial Revivalism: The Social Sources of Charismatic Authority, 1737–1770," diss., Univ. of Pennsylvania, 1966; Martin E. Lodge, "The Great Awakening in the Middle Colonies," diss., Univ. of California, Berkeley, 1964.

47. Arthur Fawcett, *The Cambuslang Revival* (London: Banner of Truth Trust, 1978).

48. *The Querist* (Philadelphia, 1740); *The Querists, to which is attached the Rev. Mr. Whitefield's Answer, the Rev. Mr. Garden's Letters, &c.* (New York: J. P. Zenger, 1740); Samuel Blair, *A Particular Consideration of a Piece, Entitled The Querist* (Philadelphia: B. Franklin, 1741). For a typical example of Whitefield's theology and its rhetorical presentation, see his *Five Sermons* (Philadelphia: B. Franklin, 1746).

49. George Gillespie, *A Sermon Against Divisions in Christ's Churches* (Philadelphia: A. & W. Bradford, 1740), pp. 2–3.

50. Synod of Philadelphia, Minutes, 31 May 1740, p. 171.

51. Presbytery of Donegal, Minutes, 7 April 1741.

52. Synod of Philadelphia, Minutes, 27 May–1 June 1741, pp. 172–74; *The Apology of the Presbytery of New Brunswick for their Dissenting from Two Acts* (Philadelphia: Benj. Franklin, 1741); Gilbert Tennent, *Remarks Upon a Protestation presented to the Synod of Philadelphia, June 1, 1741* (Philadelphia: Benj. Franklin, 1741).

53. "A Protestation presented to the Synod June 1, 1741," in Synod of Philadelphia, Minutes, pp. 186–91, citations, p. 189.

54. John Thomson, *The Government of the Church of Christ, and the Authority of Church Judicatories established on a Scripture Foundation* (Philadelphia: A. Bradford, 1741), p. iv.

55. Tennent, *Remarks Upon a Protestation*, p. 23.

56. John Gillies, *Memoirs of Rev. George Whitefield*, ed. C. E. Stowe, (Philadelphia: Leary, Getz & Co., 1859), p. 53n.

Chapter 7

1. Many historians have argued that the primary issue of the Great Awakening was clerical-lay control. While I have found this to be an important factor, I do not believe it to have been a causal one. For an intelligent exposition of the Great Awakening and anticlericalism, see Martin E. Lodge, "The Great Awakening in the Middle Colonies," diss. Univ. of California, Berkeley, 1964.

2. John Thomson, *The Doctrine of Convictions set in a Clear Light* (Philadelphia: A. Bradford, 1741).

3. Jonathan Dickinson, *A Display of God's special Grace* (Boston: Rogers & Fowle, 1742), citations, pp. i-ii, 20.

4. Dickinson, *Display of God's Grace*, citations, pp. 67, 69. See also his *Familiar Letters to a Gentleman* (Boston: Rogers & Fowle, 1745). In the preface to this pamphlet he admits the existence of emotional raptures and discountenances them as counterfeit.

5. Samuel Blair, *A Short and Faithful Narrative of the late Remarkable Revival of Religion* (Philadelphia: William Bradford [1744]), citations, pp. 27, 34. See page 27 on for discussions of communions.

6. Richard Webster, *A History of the Presbyterian Church in America, From its Origin Until the Year 1760* (Philadelphia: Joseph M. Wilson, 1857), pp. 175–76. These pages contain a summary of several of the congregational separations experienced during the Great Awakening. The details have been collected from presbytery and synod minutes, including some that are now lost.

7. "An account of the origin, progress, and present state of the Second Presbyterian Church in the city of Philadelphia, May 2, 1791," manuscript, possibly written by Samuel Hazard, descendant of some of the original church members and author of several pieces of Presbyterian history, Presbyterian Historical Society, Philadelphia.

8. Presbytery of New Brunswick, Minutes, 2 June 1741, 29 May 1742, 27

May 1743, 11–12 August 1743, 11 September 1744, Presbyterian Historical Society, Philadelphia.

9. Gilbert Tennent, manuscript list of subscribers not members of the Second Presbyterian Church of Philadelphia, Presbyterian Historical Society, Philadelphia. William Allen headed the list, while Governor Reed gave the healthy sum of thirty pounds.

10. Committee of the Second Presbyterian Church of Philadelphia, Minutes, 30 August 1760, 30 May 1765, 15 December 1769, Presbyterian Historical Society, Philadelphia. The steeple gallery was added for Negroes, usually the servants of members.

11. Synod of Philadelphia, Minutes, 2 June 1740, printed in *Minutes of the Presbyterian Church in America, 1706–1788*, ed. Guy S. Klett (Philadelphia: Presbyterian Historical Society, 1976), p. 172; Presbytery of Philadelphia, Minutes, 27 May 1740, 27 May 1741, Presbyterian Historical Society, Philadelphia.

12. Synod of Philadelphia, Minutes, 2 June 1740, p. 171.

13. Presbytery of Donegal, Minutes, 11 August 1741, Presbyterian Historical Society, Philadelphia.

14. Presbytery of Donegal, Minutes, 21 June 1743.

15. Presbytery of Donegal, Minutes, 18 June 1740, 9 April 1741, 1 August 1744; Synod of Philadelphia, Minutes, 28 May 1741, p. 173.

16. Presbytery of Suffolk, Minutes, 21 April 1747, 15–17 June 1748, 19 October 1748, 12 April 1749, 14 November 1749, 5 April 1750, 17 April 1750, 4 April 1753, Presbyterian Historical Society, Philadelphia.

17. Presbytery of New Brunswick, Minutes, 17 May 1753, 22 January 1754, 28 May 1755, 22 October 1755; Presbytery of Philadelphia, Minutes, 16 May 1758, 3 October 1759, 24 May 1760, 20 August 1760.

18. Presbytery of New Brunswick, Minutes, 4 October 1749, 16 May 1750, 17 May 1750, 7 November 1750; Synod of New York, Minutes, 27 May 1750, printed in *Minutes of the Presbyterian Church in America*, p. 273.

19. Presbytery of New Brunswick, Minutes, 2 June 1741.

20. Samuel Davies, *The State of Religion Among the Protestant Dissenters in Virginia* (Boston: S. Kneeland, 1751). The quotation itself is an account by Samuel Morris, a lay leader in Hanover, as quoted by Davies, p. 14. These details have been fleshed out from other sources in George William Pilcher, *Samuel Davies, Apostle of Dissent in Colonial Virginia* (Knoxville: Univ. of Tennessee Press, 1971), especially pp. 3–34.

21. Davies, *State of Religion in Virginia*; Presbytery of Donegal, Minutes, 1 August 1754.

22. Samuel Buell, *A Faithful Narrative of the Remarkable Revival of Religion in the Congregation of East-Hampton, Long Island* (New York: Samuel Brown, 1766). See also Caleb Smith, *A Brief Account of the Life of the Late Rev. Caleb Smith* (Woodbridge: James Parker, 1763).

23. Blair, *Short and Faithful Narrative*; Davies, *State of Religion in Virginia*; Buell, *Revival of Religion in East-Hampton*. See also John Pierson, *The Faithful*

Minister: A Funeral Sermon . . . Occasioned by the Death of the Rev. Mr. Jonathan Dickinson (New York: James Parker, 1748) in which Pierson lauds, among other qualities, Dickinson's sacramental ministry.

24. Leonard J. Trinterud, *The Forming of an American Tradition* (Philadelphia: Westminster Press, 1949), chs. 3–6, pp. 53–121.

25. The biographical information contained in both tables has been taken from the following three works: James McConnell, *Fasti of the Irish Presbyterian Church, 1613–1840*, revised by Samuel G. McConnell (Belfast: Presbyterian Historical Society, 1951); Hew Scott, *Fasti Ecclesaie Scoticanae* (Edinburgh: Tweeddale Court, 1925); and Webster, *Presbyterian Church to 1760*.

26. Synod of New York and Philadelphia, Minutes, 22 May 1758, printed in *Minutes of the Presbyterian Church*, pp. 340–43, citation, p. 341.

27. Ibid., p. 341.

28. Ibid., p. 342.

29. Dickinson, *Letters to a Gentleman*, especially pp. 182–209, 294–347. See also Dickinson, *Display of God's Grace*; Dickinson, *A Defence of the Dialogue Intitled A Display of God's special Grace* (Boston: J. Draper for S. Eliot, 1743); Samuel Blair, *The Doctrine of Predestination Truly & Fairly Stated* (Philadelphia: B. Franklin, 1742); Blair, *A Persuasive to Repentance* (Philadelphia: W. Bradford [1743]); and Gilbert Tennent, *Some Account of the Principles of the Moravians* (London: for S. Mason, 1743).

30. Gilbert Tennent to Jonathan Dickinson, 12 February 1742, as cited in Webster, *Presbyterian Church to 1760*, pp. 189–91; originally published in the *Pennsylvania Gazette*.

31. William Howland Kenney, "George Whitefield and Colonial Revivalism: The Social Sources of Charismatic Authority, 1737–1770" diss., Univ. of Pennsylvania, 1966, ch. 4.

32. Samuel Blair, *Animadversions on the Reasons of Mr. Alex. Creaghead's Seceding from the Judicatures of this Church* (Philadelphia: William Bradford [1743]).

33. Gilbert Tennent, *The Danger of Spiritual Pride represented,* preached at Philadelphia, 30 December 1744 (Philadelphia: William Bradford [1745]), citations, pp. 19, 14.

34. Gilbert Tennent, *The Necessity of studying to be quiet, and doing our own Business* (Philadelphia: William Bradford, 1744); G. Tennent, *Brotherly Love recommended, by Argument of the Love of Christ* (Philadelphia: Benjamin Franklin and David Hall, 1748). See also G. Tennent, *A Persuasive to the Right Use of the Passions in Religion* (Philadelphia: W. Delap, 1760).

35. Dickinson, *Display of God's Grace*; Samuel Davies, *A Sermon on Man's Primitive State and the First Covenant* (Philadelphia: William Bradford, 1748).

36. Tennent to Dickinson, 12 February 1742.

37. See also Lodge, "The Great Awakening," pp. 258–74, for other details of the "disintegration" of the awakening among the Presbyterians.

38. Synod of New York and Philadelphia, Minutes, 22 May 1758, p. 342.

39. For sermons that explicitly outline this conversion model, see Jonathan Dickinson, *A Call to the Weary & heavy Laden to come unto Christ for Rest* (New York: Wm Bradford, 1740); Dickinson, *The Witness of the Spirit* (Boston: S. Kneeland and T. Green, 1740); Samuel Finley, *Clear Light put out in obscure darkness* (Philadelphia: B. Franklin, 1743); Blair, *A perswasive to Repentance*; and Gilbert Tennent, *Love to Christ a necessary Qualification in Order to feed his Sheep* (Philadelphia: William Bradford, 1744).

Bibliography

I. Primary Sources

A. Manuscripts

Abbington Presbytery. Records of the Abbington Presbytery, 1752–58. Original MSS, Presbyterian Historical Society, Philadelphia.

Aghadowey Congregation. Commonplace Book, c. 1700. Original MS, Presbyterian Historical Society, Belfast, located at Union Theological College, Belfast, Box 2.

"An Account of the Origin, Progress, and Present State of the Second Presbyterian Church." MS copy (1792), Presbyterian Historical Society, Philadelphia.

Andrews, Jedidiah. Letters. MS copies, Presbyterian Historical Society, Philadelphia.

Antrim Presbytery. Minutes of the Antrim Meeting, 1654–58; 1671–75; 1683–91. Handwritten transcript by W. T. Latimer, (1900 and 1901), examined and corrected by John S. MacMaster (1935), Public Record Office of Northern Ireland, Belfast, D/1759/1A/1.

Associate Presbytery. Minutes of the Associate Presbytery, 1733–47. MS copy, 2 vols., certified by James Hall (1824), Scottish Record Office, Edinburgh, CH3/27/1, CH3/28/1.

Auchterarder Presbytery. Minutes of the Auchterarder Presbytery, 1668–87; 1703–45. Original MSS, 5 vols., Scottish Record Office, Edinburgh, CH2/619/1, CH2/619/26–29.

Aughnacloy Congregation. Committee Book, 1743–82. Original MSS, Presbyterian Historical Society, Belfast, located at Union Theological College, Belfast, Box 12.

Ballybay Congregation. Minute Book kept by the Congregation and Eldership of the new Meeting House of Ballybay, 1751–67. MSS copy, Presbyterian Historical Society, Belfast, located at Union Theological College, Belfast, Box 5.

Burt Congregation. Minutes of Burt Kirk Session, 1676–1719. Original MSS,

241

Presbyterian Historical Society, Belfast, located at Union Theological College, Belfast, on shelf.

Cambuslang Congregation. Minutes of the Kirk Session of Cambuslang, 1658–1748. Original MSS, 2 vols., Scottish Record Office, Edinburgh, CH2/415/1–2.

Carluke Congregation. Minutes of Carluke Kirk Session, 1645–62; 1694–1750. Original MSS, 3 vols., Scottish Record Office, Edinburgh, CH2/56/1–3.

Cook, John. Diary, 1698–1705. MS copy (late eighteenth/early nineteenth century), Presbyterian Historical Society, Belfast, located at Union Theological College, Belfast, Box 1.

Dalmellington Congregation. Session Minutes of Dalmellington, 1641–62; 1691–1720; 1728–39. Original MSS, 2 vols., Scottish Record Office, Edinburgh, CH2/85/12.

Donegal Presbytery. Minutes and Proceedings of the Presbytery of Dunagal, 1732–50; 1759–69. Original MSS, 2 vols., Presbyterian Historical Society, Philadelphia.

Douglas, Colonel James. Letters to Colonal James Douglas, 1682–88. Original MSS, National Library of Scotland, Edinburgh, MS 214.

Dron Congregation. Session Records of Dron, 1632–1750. Original MSS, 3 vols., Scottish Record Office, Edinburgh, CH2/93/1–3.

Dundonald Congregation. Session Records of Dundonald, 1602–43; 1702–31; 1731–81. Original MSS, 3 vols., Scottish Record Office, Edinburgh, CH2/104/12.

Duchal, [James]. Two Sermons delivered at Comber, 21 April 1723 and 16 August 1730. Original MSS, Presbyterian Historical Society, Belfast, located at Union Theological College, Belfast, Box 1.

Duchal, James. Correspondence between James Duchal and Thomas Drennan, 1739–59. Original MSS, Presbyterian Historical Society, Belfast, located at Union Theological College, Belfast, Box 9.

Glasgow and Ayr Synod. Minutes of the Synod of Glasgow and Ayr, 1687–1760. Original MSS, 3 vols., Scottish Record Office, Edinburgh, CH2/464/1–3.

Gordon, Robert. Diary, 1707–43. MS copy, Presbyterian Historical Society, Belfast, located at Union Theological College, Belfast, Box 10.

Hamilton Presbytery. Minutes of Hamilton Presbytery, 1687–1757. Original MSS, 3 vols., Scottish Record Office, Edinburgh, CH2/393/1–3.

Kennedy, J. Diary, 1714–37. Original MSS, Presbyterian Historical Society, Belfast, located at Union Theological College, Belfast, Box 12.

Killyleagh Presbytery. Minutes of Killyleagh Presbytery, 1725–32. Original MSS, Presbyterian Historical Society, Belfast, located at Union Theological College, Belfast, Box 3.

Kirkdonald Congregation. Session Book of Kirkdonald, 1678–1713. Original MSS, Presbyterian Historical Society, Belfast, located at Union Theological College, Belfast, Box 6.

Kirkoswald Congregation. Session Records of Kirkoswald, 1619–53; 1694–1755. Original MSS, 2 vols., Scottish Record Office, Edinburgh, CH2/562/1, 6.

Laggan Presbytery. Minutes of the Laggan Meeting, 1672–1700. Original MSS, 2 vols., Presbyterian Historical Society, Belfast, located at Union Theological College, Belfast, on shelf.

Lanark Presbytery. Minutes of the Presbytery of Lanark, 1623–1749. Original MSS, 8 vols., Scottish Record Office, Edinburgh, CH2/234/1–8.

Larne Congregation. Minutes of Larne Session, 1699–1701. MS copy, Presbyterian Historical Society, Belfast, located at Union Theological College, Belfast, Box 9.

Lauderdale Papers. Correspondence to the Duke of Lauderdale from James Sharp, Archbishop of St. Andrews, and Alexander Burnet, Archbishop of Glasgow and later of St. Andrews, 1660–81. Original MSS, National Library of Scotland, Edinburgh, MS 2512.

Lee Papers. Papers of the very Rev. John Lee, 1637–1800. Original MSS, National Library of Scotland, Edinburgh, MSS 3430 and 3431.

Lewistown Presbytery. Minutes and Transactions of Lewistown Presbytery, 1758–74. Original MSS, Presbyterian Historical Society, Philadelphia.

Lisburn Congregation. Minutes of the Lisburn Session, 1711–30 complete, 1730–63 incomplete. Original MSS, 3 vols., Presbyterian Historical Society, Belfast, located at Union Theological College, Belfast, Box 11.

Londonderry Synod. Minutes of the Sub-Synod of Londonderry, 1706–37; 1744–1802. Original MSS, 2 vols., Presbyterian Historical Society, Belfast, located at Union Theological College, Belfast, Box 4.

McBryde, John. Daniel M——'s notes on Sermons by John McBryde and others, delivered in Belfast, 1704. Original MSS, Presbyterian Historical Society, Belfast, located at Union Theological College, Belfast, Box 1.

McKenzie, John. Sermons delivered during the Seige of Derry, 1681. Original MSS (notes), Presbyterian Historical Society, Belfast, located at Union Theological College, Belfast, Box 1.

Martin, John. Twenty-six sermons delivered at Antrim, 1746–56. Original MSS, Presbyterian Historical Society, Belfast, located at Union Theological College, Belfast, Box 6.

Maxwell, Sir George, of Pollack. Diary, 1655–66. Original MS, National Library of Scotland, Edinburgh, MS 3150.

Monaghan Presbytery. Minutes of the Monaghan Presbytery, 1702–12. Original MSS, Presbyterian Historical Society, Belfast, located at Union Theological College, Belfast, Box 10.

New Brunswick Presbytery. Records of New Brunswick Presbytery, 1738–71. Original MSS, 2 vols., Presbyterian Historical Society, Philadelphia.

New Castle Presbytery. Records of the New Castle Presbytery, 1716–31. Original MSS, Presbyterian Historical Society, Philadelphia.

Paisley Presbytery. Minutes of the Presbytery of Paisley, 1626–1752. Original MSS, 8 vols., Scottish Record Office, Edinburgh, CR2/294/1–8.

Philadelphia Presbytery. Minutes of the Presbytery of Philadelphia, 1733–46; 1758–90. Original MSS, 2 vols., Presbyterian Historical Society, Philadelphia.

Philadelphia, First Presbyterian Church. Correspondence and Papers, 1749–1800. Original MSS, Presbyterian Historical Society, Philadelphia.

———. Minutes of the meetings of the Congregation and Committee, 1747–72. Original MSS, Presbyterian Historical Society, Philadelphia.

Philadelphia, Second Presbyterian Church. Congregation, Corporation, and Trustees Minutes, 1749–80. Original MSS, Presbyterian Historical Society, Philadelphia.

———. Session Minutes, 1747–62. Original MSS, Presbyterian Historical Society, Philadelphia.

Route Presbytery. Minutes of the Route Presbytery, 1701–06. Original MSS, Presbyterian Historical Society, Belfast, located at Union Theological College, Belfast, Box 1.

Sempil, Gabriel. Autobiography [c. 1665]. MS copy (1728), National Library of Scotland, Edinburgh, MS 5746.

St. John's Congregation, Perth. Minutes of the Session of St. John's, Perth, 1577–1612, 1615–24, 1631–34, 1665–86, 1692–1745. Original MSS, 17 vols., Scottish Record Office, Edinburgh, CH2/521/1–17.

Suffolk Presbytery. Records of the Suffolk Presbytery, 1747–56. Original MSS, Presbyterian Historical Society, Philadelphia.

Strabane Presbytery. Minutes of the Presbytery of Strabane, 1717–40. Original MSS, Presbyterian Historical Society, Belfast, located at Union Theological College, Belfast, on shelf.

Templepatrick Congregation. Minutes of the Session of Templepatrick Church, 1646–1743. Original MSS, 6 vols., Public Record Office of Northern Ireland, Belfast, CR4/12B/1–6.

Wodrow, Robert. Correspondence: Irish Letters, 1705–26. MS copy, Presbyterian Historical Society, Belfast, located at Union Theological College, Belfast, Box 4.

B. Published Editions of Manuscript Sources

Adair, Patrick. *A True Narrative . . . of the Presbyterian Church in Ireland (1623–1670).* Introd. and notes by W. D. Killen. Belfast: C. Aitchison, 1866. The following historical account was appended to Adair's piece: Stewart, Andrew, "History of the Church in Ireland," pp. 213–321.

Baillie, Robert. *The Letters and Journals of Robert Baillie.* Ed. David Laing. 3 vols. Edinburgh, n.p., 1841.

Blair, Robert. *The Life of Mr. Robert Blair . . . his Autobiography from 1593–1636 with Supplement to his Life, and Continuation of the History of the Times to 1680, by . . . Mr. William Row.* Ed. Thomas M'Crie. Edinburgh: The Wodrow Society, 1848.

Franklin, Benjamin. *Benjamin Franklin's Autobiography*. Ed. J. A. Leo LeMay
and P. M. Zall. New York: Norton, 1986.

Gillies, John. *Historical Collections Relating to Remarkable Periods of the Success
of the Gospel* [1754]. Kelso, Scotland: John Rutherford, 1845.

———. *Memoirs of George Whitefield*. Philadelphia: Leary, Getz & Co., 1859.

Klett, Guy S., ed. *Minutes of the Presbyterian Church in America 1706–1788*.
Philadelphia: Presbyterian Historical Society, 1976.

Livingstone, John. *A Brief Historical Relation of the Life of Mr. John Livingstone
. . . written by himself*. Ed. Thomas Houston. Edinburgh: John Johnston,
1848.

Makemie, Francis. *The Life and Writings of Francis Makemie*. Ed. Boyd Schlenther. Philadelphia: Presbyterian Historical Society, 1971.

Miscellaneous Manuscripts on Eighteenth Century Ulster Emigration to North
America. Education Facsimiles 121–40, Public Record Office of Northern
Ireland, Belfast.

Mitchell, Alex F. and Struthers, John, eds. *Minutes of the Sessions of the Westminster Assembly of Divines*. Edinburgh: W. Blackwood & Sons, 1974.

Ulster Synod. *Records of the General Synod of Ulster, 1691–1820*. 3 vols. Belfast:
John Reid and Co., 1898.

Whitefield, George. *Journals* [1738–47]. London: Banner of Truth Trust, 1960.

———. *Letters of George Whitefield for the period 1734–1742* [1771]. London:
Banner of Truth Trust, 1976.

Wodrow, Robert. *The History of the Sufferings of the Church of Scotland from the
Restoration to the Revolution*. 4 vols. Glasgow: Blackie & Son, 1836.

Woodmason, Charles. *The Carolina Backcountry on the Eve of the Revolution*. Ed.
and Introd. by Richard J. Hooker. Chapel Hill: Univ. of North Carolina Press,
1953.

C. Pamphlets

Abernethy, John. *A Defence of the Seasonable Advice*. Belfast: James Blow, 1724.

———. *A Sermon Recommending the Study of Scripture Prophecies*. Preached
before the Synod of Ulster, 19 June 1716. Belfast: James Blow, 1716.

———. *Persecution contrary to Christianity. Sermon delivered at Wood Street,
Dublin, 23 October 1735 (anniversary of the Irish Rebellion)*. Dublin: S.
Powell, 1735.

———. *Religious Obedience founded on Personal Persuasion. A Sermon Preach'd
at Belfast the 9th of December, 1719*. Belfast: James Blow, 1720.

———. *Seasonable Advice to the Protestant Dissenters in the North of Ireland*.
Preface by Nath. Weld, J. Boyse, and R. Choppin. Dublin: James Carson,
1722.

A Friendly Advice for Preserving the Purity of Doctrine and Peace of the Church.
Edinburgh: J. Mossman and Company, 1722.

A Letter concerning the Defections, Sins, and Backslidings of the Church of Scotland. N.p. 1720.

A Letter from a Gentleman in Scotland to His Friend in New England. Boston: T. Fleet, 1743.

A Letter to a Gentleman at Edinburgh, a Ruling Elder of the Church of Scotland, concerning Proceedings of the last General Assembly. N.p. [1721].

A Letter to the Author of the Antinomianism of the Marrow of Modern Divinity detected. Edinburgh, 1722.

Alison, Francis. *Peace and Union Recommended.* Preface by Gilbert Tennent. Philadelphia: W. Dunlap, 1758.

An Account of Several Things. London, 1719.

An Account of the late Proceedings of the Dissenting Ministers at Salters Hall in a letter to the Rev^d Dr. Gale. London, 1719.

An Apology for the English Presbyterians. London, 1699.

An Appendix to Mr. Robe's Historical and Remarking Paper: vindicating the Late Act of the Assembly concerning Mr. Leechman's Affair. Edinburgh [between 1770 and 1774].

An Essay upon Gospel and Legal Preaching. Edinburgh: James Davidson, 1723.

An Extract of the Minutes of the Commission of the Synod, Relating to the Affair of the Reverend Mr. Samuel Hemphill. Philadelphia: Andrew Bradford, 1735.

An Impartial State of the Late Differences Amongst the Protestant Dissenting Ministers at Salters-Hall. London, 1719.

Antrim Presbytery. A Letter from the Presbytery of Antrim to the Congregations under their Care: Occasioned by the Uncharitable Breach. Belfast: James Blow, 1726.

Apology of the Presbytery of New Brunswick for their Dissenting from Two Acts. Philadelphia: Benj. Franklin, 1741.

A Publick Testimony. Edinburgh: Thomas Lumisden and John Robertson, 1732.

A Short Account of the Remarkable Conversions at CAMBUSLANG. Glasgow: Robt. Smith, 1742.

A Short Account of the Remarkable Conversions at Cambuslang. Glasgow: Robt. Smith, 1742.

A Sober Defence of the Reverend Ministers who, by a Subscription, have lately Declar'd their Faith in the Trinity. London, 1719.

Associate Presbytery. *Act of the Associate Presbytery.* N.p., 1739.

Associate Presbytery. *Act, Declaration, and Testimony for the Doctrine, Worship, Discipline, and Government of the Church of Scotland.* Edinburgh: Thomas Lumisden and John Robertson, 1737.

A True Relation of some Proceedings at Salters-Hall. London, n.d.

A Vindication of the Doctrine of Grace from the Charge of Antinomianism. Edinburgh: Robert Brown, 1718.

A Vindication of the Reverend Commission of the Synod: In Answer to Some Observations on their Proceedings against the Reverend Mr. Hemphill. Philadelphia: Andrew Bradford, 1735.

Blair, John. *Animadversions on a Pamphlet*. Philadelphia: William and Thomas Bradford, 1766.

———. *The Successful Preacher* (incomplete copy). Philadelphia: William Bradford, 1761.

Blair, Samuel. *Animadversions on the Reasons of Mr. Alex. Creaghead's Seceding from the Judicatures of this Church*. Philadelphia: William Bradford [1743].

———. *A particular Consideration of A Piece Entitled the Querists*. Philadelphia: B. Franklin, 1741.

———. *A perswasive to Repentance . . . preached at Philadelphia, 1739*. Philadelphia: W. Bradford [1743].

———. *A Short and Faithful Narrative of the late Remarkable Revival of Religion*. Philadelphia: William Bradford [1744].

———. *A Vindication of the Brethren . . . against the Charges of the Rev. Mr. John Thompson*. Philadelphia: B. Franklin, 1744.

———. *The Doctrine of Predestination*. Philadelphia: B. Franklin, 1742.

———. *The Gospel Method of Salvation*. New York: William Bradford, 1737.

Bostwick, David. *Self disclaim'd and Christ exalted*. Printed following Alison, Francis, *Peace and Union Recommended*. Philadelphia: W. Dunlap, 1758.

Boyse, J[oseph]. *A Vindication of a Private Letter*. Belfast: James Blow, 1726.

———. *A Vindication of the True Deity of our Blessed Saviour in Answer to a Pamphlet Entitled An Humble Inquiry into the Scripture-account of Jesus Christ, &c.* Dublin, 1703.

Bradbury, Thomas. *An Answer to the Reproaches Cast on those Dissenting Ministers Who Subscrib'd their Belief of the Eternal Trinity*. London, 1719.

Bruce, Michael. *The Rattling of the Dry Bones. Preached at Carluke, May 1672.* Printed in a volume with four other sermons. N.p. [1672].

Buell, Samuel. *A Faithful Narrative of the Remarkable Revival of Religion in the Congregation of East-Hampton, on Long Island*. New York: Samuel Brown, 1766.

Burr, Aaron. *The Watchman's Answer to the Question What of the Night, etc.* 3rd ed. Boston: S. Kneeland, 1757.

Calamy, Edmund. *Comfort and Counsel to Protestant Dissenters*. London, 1712.

Carlisle, John. *The Nature of Religious Zeal. Sermon preached at the General Synod of Ulster at Antrim, 18 June 1745.* Belfast: James Magee, 1745.

Cherry, George. *Sermon Preached before the Particular Synod of Ardmagh on Tuesday, July 29, 1736.* Dublin: M. Rhames, 1736.

Clerk, Matthew. *A Letter from the Belfast Society to the Rev. Matthew Clerk . . . with an Answer to the Society's Remarks*. [Belfast] 1723.

Craighead, Robert, Sr. *Advice to Communicants*. Glasgow: Robert Sanders, 1714.

———. *A Plea for Peace, or, the Nature, Causes, Mischief, and Remedy of Church-Divisions. Sermon preach'd at Belfast 22 June 1720.* Dublin: George Grierson, 1721.

Craighead, Robert [Jr.]. *The True Terms of Christian and Ministerial Communion founded on Scripture alone. A sermon by the late Robert Craighead.* Preface by John Abernethy. Dublin: S. Powell, 1739.

[Dalrymple, David]. *An Account of Lay Patronage in Scotland.* Edinburgh: J. M., 1712.

Davies, Samuel. *A Sermon on Man's Primitive State and the First Covenant.* Philadelphia: William Bradford, 1748.

————. *The State of Religion among the Protestant Dissenters in Virginia.* Boston: S. Kneeland, 1751.

Delap, Samuel. *Remarks on Some Articles of the Seceders New Covenant.* Originally published in Belfast, 1754. Rpt. Lancaster: W. Dunlap, 1754.

Dickinson, Jonathan. *A Ca'l to the Weary & heavy Laden to come unto Christ for Rest.* New York: Wm Bradford, 1740.

————. *A Defence of a Sermon Preached at Newark [Inti]tuled, the Vanity of Human Institutions in the Worship of God.* New York: J. Peter Zenger, [1737].

————. *A Defence of Presbyterian Ordination.* Boston: for Daniel Henchman, 1724.

————. *A Defence of the Dialogue Intitled A Display of God's special Grace.* Boston: J. Draper for S. Eliot, 1743.

————. *A Display of God's Special Grace in A familiar Dialogue.* Boston: Rogers and Fowle for S. Eliot, 1742.

————. *Familiar Letters to a Gentleman.* Boston: Rogers & Fowle and J. Blanchard, 1745.

————. *Remark's on Mr. Gales Reflections.* New York: T. Wood, 1721.

————. *Remarks Upon A Discourse intituled An Overture Presented to the Reverend Synod of Dissenting Ministers sitting in Philadelphia, in the Month of September, 1728.* New York: J. Peter Zenger, 1729.

————. *Remarks Upon a Pamphlet.* Philadelphia: Andrew Bradford, 1735.

————. *Remarks Upon the Postscript to the Defence. . . .* Boston: for D. Henchman, 1724.

————. *Sermon preached at the Opening of the Synod at Philadelphia September 19, 1722.* Boston: T. Fleet for S. Gerrish, 1723.

————. *The Danger of Schisms and Contentions.* New York: J. Peter Zenger, 1739.

————. *The Nature and Necessity of Regeneration.* New York: James Parker, 1743.

————. *The Vanity of Human Institutions in the Worship of God.* New York: John Peter Zenger, 1736.

————. *The Witness of the Spirit.* Boston: S. Kneeland & T. Green, 1740.

Drake, Roger. *A Boundary to the Holy Mount, or, A Barre against Free Admission to the Lord's Supper.* London: Abraham Miller, 1653.

Emlyn, Thomas. *A True Narrative of the Proceedings of the Dissenting Ministers of Dublin against Mr. Thomas Emlyn.* London: J. Darby, 1719.

[Erskine, Ralph]. *A Review of an Essay upon Gospel and Legal Preaching*. Edinburgh: Ja: McEuen, 1723.

Evans, David. *Law and Gospel or Man wholly ruined by the Law and Recovered by the Gospel*. Philadelphia: B. Franklin and D. Hall, 1748.

————. *The Minister of Christ and the Duties of his Flock*. Philadelphia: B. Franklin, 1732.

Ferguson, Victor. *A Vindication of the Presbyterian Ministers in the North of Ireland: Subscribers and Non-subscribers*. Belfast: James Blow, 1721. Ferguson was actually the publisher and not the author. The pamphlet was published anonymously. Historians now believe the author was either John Abernethy or James Kirkpatrick.

Finley, Samuel. *A Letter to a Friend*. [Boston, 1745.]

————. *Christ Triumphing and Satan Raging*. Philadelphia: B. Franklin, 1741.

————. *Clear Light put out in obscure Darkness*. Philadelphia: B. Franklin, 1743.

————. *The successful Minister of Christ distinguished in Glory*. Philadelphia: William Bradford, 1764.

Fisher, Edward. *The Marrow of Modern Divinity*. Annotated by Thomas Boston. Glasgow: John Bryce and David Patterson, 1722. *The Marrow of Modern Divinity* was first published in 1645.

Fisher, James. *A Review of the Preface to a Narrative. . . .* Glasgow: J. Bryce, 1742.

Fleming, Robert. *The Fulfilling of the Scripture* [1743]; rpt. Charleston: Samuel Etheridge, 1806.

[Franklin, Benjamin]. *A Defense of the Rev. Mr. Hemphill's Reservations*. Philadelphia: B. Franklin, 1735.

————. *A Letter to a Friend in the Country Containing the Substance of a Sermon Preach'd at Philadelphia, in the Congregation of the Rev. Mr. Hemphill*. Philadelphia: B. Franklin, 1735.

————. *Some Observations on the Proceedings Against the Rev. Mr. Hemphill; with a Vindication of his Sermons*. 2nd Ed. Philadelphia: B. Franklin, 1735.

General Assembly of the Church of Scotland. *Act . . . concerning the Method of Planting Vacant Congregations*. Edinburgh: J. Davidson & R. Fleming, 1732.

General Assembly of the Church of Scotland. *Act . . . concerning the Ministers seceding from the said Church*. N.p. [1738].

General Assembly of the Church of Scotland. *Copy of a Libel against Messrs. Ebenezer Erskine and Others, Ministers, who have seceded from the Church of Scotland*. N.p., 1739.

Gillespy [Gillespie], George. *Remarks upon Mr. George Whitefield*. Philadelphia: B. Franklin for author, 1744.

————. *A Sermon Against Divisions in Christ's Churches*. Philadelphia: A. and W. Bradford, 1740.

————. *A Treatise Against the Deists or Free-Thinkers: Proving the Necessity of Revealed Religion*. Philadelphia: A. Bradford, 1735.

Gowan, Thomas. *The Necessity of standing fast by our Christian Liberty.* Sermon preached at Lisburn 28 March 1714. Belfast: Robert Gardner, n.d.

Haliday, Samuel. *A Letter to the Reverend Mr. Gilbert Kennedy.* Belfast: James Blow, 1725.

————. *Reasons Against the Imposition of Subscription to the Westminster Confession of Faith.* Belfast: James Blow, 1724.

[Hemphill, Samuel]. *Some General Remarks Argumentative and Historical on the Vindication Publish'd by Dr. Ferguson.* N.p., 1722.

Higginbotham, Robert. *Reasons Against the Overtures.* . . . Belfast: James Blow, 1726.

Kennedy, Gilbert. *A Defence of the Principles and Conduct of the Rev'd General Synod of Ulster.* Belfast: Robert Gardner, 1724.

————. *The Great Blessing of Peace and Truth in Our Days.* A Sermon preach'd at Belfast on Tuesday, April 25, 1749. Belfast, 1749.

[Kirkpatrick, James.] *An Essay upon the Important Question, Whether there be a Legislative, Proper Authority in the Church . . . By Some Non-Subscribing Ministers in the North of Ireland.* Belfast: James Blow, 1737.

————. *A Scriptural Plea Against a Fatal Rupture and Breach of Christian Communion.* Belfast, 1724.

McBride, Robert. *The Overtures transmitted by the General Synod, 1725, Set in a Fair Light.* Belfast: Robert Gardner, 1726.

Masterton, C. *Christian Liberty founded in Gospel Truth.* Belfast: Robert Gardner, 1725.

Morgan, Joseph. *A Discourse shewing How the Nature of Sin is in itself A Misery to the Sinner.* Boston: Gamaliel Rogers for Samuel Gerrish, 1728.

————. *A Letter to the Authors of a Discourse intitled Some Short Observations.* New London, Conn., 1724.

————. *Love to our Neighbours Recommended, and the Duties thereof Importunately Verged.* New London, Conn.: T. Green, 1727.

————. *The Great Concernment of Gospel Ordinances.* New York, 1712.

————. *The Nature and Original of Sin Explained, and the Fruits & Effects Bewailed.* New London, Conn.: T. Green, 1727.

————. *The Only Effectual Remedy Against Mortal Errors.* New London, Conn.: T. Green, 1725.

Narrative of the Proceedings of Seven General Synods . . . in which they issued a Synodical Breach. Belfast, 1727.

Nevin, Thomas. *The Tryal of Thomas Nevin.* Belfast: James Blow, 1725.

Pemberton, Ebenezer. *A Sermon Preach'd at the Ordination of the Reverend Mr. Walter Wilmot.* Boston: J. Draper for D. Henchman, 1738.

————. *Sermon preach'd before the Commission of the Synod at Philadelphia April 20, 1735.* New York: John Peter Zenger, 1735.

Pierce, James. *A Defence of the Case of the Ministers Ejected at Exon.* London, 1719.

————. *A Letter to a Subscribing Minister.* London, 1719.

————. *The Case of the Ministers Ejected at Exon.* London, 1719.

Pierson, John. *The Faithful Minister: A Funeral Sermon . . . Occasioned by the Death of the Rev. Mr. Jonathan Dickinson.* New York: James Parker, 1748.

Presbyterian Church in the U.S.A. *A Declaration of the Presbyteries of New-Brunswick and New-Castle.* Philadelphia: William Bradford, 1743.

Prince, Thomas, Ed. *The Christian History.* Boston: S. Kneeland and T. Green, 1744–45.

Protestation of Several Ministers of the Gospel against the General Assembly's illegal Proceedings, upon the Head of Doctrine. N.p., 1722.

Queries Agreed unto by the Commission of the General Assembly . . . Together with the Answers Given. N.p., 1722.

Queries to the Friendly Adviser. N.p., 1722.

Remarks upon the Answers of the Brethren to the Queries. Edinburgh, 1722.

Remarks upon the Representation Made by the Kirk of Scotland Concerning Patronage. N.p. [1712].

Robe, James. *A Faithful Narrative of the Extraordinary Work of the Spirit of God at Kilsyth.* Edinburgh, 1742.

————. *Mr. Robe's first Letter to the Rev.ᵈ Mr. James Fisher.* Glasgow, 1742.

————. *Mr. Robe's Fourth Letter to the Reverend Mr. James Fisher.* Glasgow, 1743.

————. *Mr. Robe's Second Letter to the Reverend Mr. James Fisher.* Glasgow, 1742.

————. *Mr. Robe's Third Letter to the Reverend Mr. James Fisher.* Edinburgh: T. Lumisden and J. Robertson, 1742.

Sermons on Sacramental Occasions by divers Ministers. Boston: J. Draper for D. Henchman, 1739.

Seward, William. *Journal of a Voyage from Savannah to Philadelphia, and from Philadelphia to England, M,DCC.XL.* London, 1740.

Short History of the Sabbath. Edinburgh: Andrew Anderson, 1705.

Smith, Caleb. *A Brief Account of the Life of the Late Rev. Caleb Smith, A.M.* Woodbridge, N.J.: James Parker, 1763.

Some Observations on the Rev. Mr. Whitefield and His Opposers. Boston: for D. Henchman, 1740.

Some Observations upon the Answers of the Brethren to the Queries. Edinburgh, 1722.

Some Short Observations made on the Presbyterian Doctrine of Election & Reprobation. Philadelphia: Andrew Bradford, 1721.

Tennent, Gilbert. *A Discourse Upon Christ's Kingly Office.* Preached September 1740, Nottingham [Pennsylvania]. Boston: for C. Harrison, 1741.

————. *A Funeral Sermon Occasion'd by the Death of the Reverend Mr. John Rowland.* Philadelphia: William Bradford, 1745.

————. *All Things come alike to All.* Preached at Philadelphia, 28 July 1745. Philadelphia: William Bradford, 1745.

————. *A Persuasive to the Right Use of the Passions in Religion.* Philadelphia: W. Dunlap, 1760.

————. *A Sermon Upon Justification.* Philadelphia: Benjamin Franklin, 1741.

————. *A Solemn Warning to the Secure World from the God of terrible Majesty.* Boston: S. Kneeland and T. Green for D. Henchman, 1735.

————. *Brotherly Love recommended, by Argument of the Love of Christ.* Philadelphia: Benjamin Franklin and David Hall, 1748.

————. *Love to Christ a necessary Qualification in Order to feed his Sheep.* Preached at Neshaminy, Dec. 14, 1743. Philadelphia: William Bradford, 1744.

————. *Remarks Upon A Protestation Presented to the Synod of Philadelphia June, 1741.* Philadelphia: Benj. Franklin, 1741.

————. *Some Account of the Principles of the Moravians.* London: for S. Mason, 1743.

————. *The Danger of an Unconverted Ministry.* Philadelphia: Benjamin Franklin, 1740.

————. *The Danger of forgetting God, describ'd, and The Duty of considering our Ways explain'd.* New York: John Peter Zenger, 1735.

————. *The Danger of Spiritual Pride represented.* Preached at Philadelphia, 30 December 1744. Philadelphia: William Bradford [1745].

————. *The Espousals or A Passionate Perswasive to a Marriage with the Lamb of God, wherein the Sinners Misery and the Redeemers Glory is Unvailed in.* New York: J. Peter Zenger, 1735.

————. *The Necessity of Receiving the Truth in Love.* New York: John Peter Zenger, 1735.

————. *The Necessity of Religious Violence in Order to Obtain Durable Happiness.* New York: William Bradford [1735].

————. *The Necessity of studying to be quiet, and doing our own Business.* Philadelphia: William Bradford, 1744.

————. *The Righteousness of the Scribes and Pharisees.* Preached Tuesday evening lecture in Boston Jan. 27, 1740.1. Boston: J. Draper for D. Henchman, 1741.

————. *The Terrors of the Lord.* Philadelphia: William Bradford, 1749.

————. *Twenty Three Sermons.* Philadelphia: William Bradford, 1744.

Tennent, John. *The Nature of Regeneration Opened, And its Absolute Necessity, In Order to Salvation Demonstrated. Also the Nature of Adoption, With its Consequent Priviledges Explained.* Printed after Tennent, Gilbert. *A Solemn Warning to the Secure World.* Boston: S. Kneeland and T. Green for D. Henchman, 1735.

The Anatomy of the Heretical Synod of Dissenters at Salters-Hall. 2nd Ed. London, 1719.

The Description of a Presbyterian. N.p., 1710.

The Querist, or, An Extract of Sundry Passages taken out of Mr. Whitefield's printed Sermons, Journals and Letters. Philadelphia, 1740.

The Querists, to which is attached the Rev. Mr. Whitefield's Answer, the Rev. Mr. Garden's Letters, etc. N.p., n.d.

The Synod. London, 1719.

Thompson, John. *An Overture Presented to the Reverend Synod of Dissenting Ministers Sitting in Philadelphia, in the Month of September, 1728.* [Philadelphia]: for author [1729].

———. *The Doctrine of Conviction set in a Clear Light.* Philadelphia: A. Bradford, 1741.

———. *The Government of the Church of Christ.* Philadelphia: A. Bradford, 1741.

Three Letters to the Reverend Mr. George Whitefield. Philadelphia: Andrew Bradford, 1739.

Tomkins, Martin. *The Case of Mr. Martin Tomkins.* London, 1719.

Webster, Alexander. *Divine Influence the True Spring of the Extraordinary Work at Cambuslang.* Edinburgh: T. Lumisden and J. Robertson, 1742.

Whitefield, George. *A Continuation of the Reverend Mr. Whitefield's Journal from his leaving New England.* Boston: G. Rogers for J. Edwards and S. Eliot, 1741.

———. *Five Sermons on the Following Subjects.* Philadelphia: B. Franklin, 1746.

Willison, John. *A Letter from Mr. John Willison . . . to Mr. James Fisher.* Edinburgh: T. Lumisden and J. Robertson, 1743.

[Witherspoon, John T.] *A Letter from a Blacksmith to the Ministers and Elders of the Church of Scotland.* 5th Ed. Dublin: George Faulkner, 1760.

II. Secondary Sources

Ahlstrom, Sidney. *A Religious History of the American People.* New Haven: Yale Univ. Press, 1972. Parts 3 and 4, pp. 261–511.

Alexander, Archibald. *The Log College.* 1851; rpt. London: Banner of Truth Trust, 1968.

Allen, Robert. "The Principle of Non-Subscription to Creeds and Confessions of Faith as exemplified in Irish Presbyterian History." Diss., The Queen's University, Belfast, 1944.

Bailie, W. D. *The Six Mile Water Revival of 1625.* Newcastle, County Down: Mourne Observer Press, 1976.

Barkley, John M. *A Short History of the Presbyterian Church in Ireland.* Belfast: Publications Board, Presbyterian Church in Ireland, 1959.

Beckett, J. C. *Protestant Dissent in Ireland, 1687–1780.* London: Faber & Faber, 1946.

Bolane, C. Gordon; Goring, Jeremy; Short, H. L.; and Thomas, Roger. *The English Presbyterians.* London: George Allen & Unwin, 1968.

Brown, Andrew William Godfrey. "Irish Presbyterian Theology in the Early Eighteenth Century." Diss., The Queen's University, Belfast, 1977.

Burleigh, J. H. S. *A Church History of Scotland*. London: Oxford Univ. Press, 1960.

Bushman, Richard L., ed. *The Great Awakening: Documents on the Revival of Religion, 1740–1745*. New York: Norton, 1970.

Butler, Jon. "Enthusiasm Described and Decried: The Great Awakening as Interpretive Fiction." *Journal of American History*, 69 (1982), 305–25.

————. *Power, Authority, and the Origins of American Denominational Order*. Philadelphia: American Philosophical Society, 1978.

Campbell, R. H., and Skinner, Andrew S., eds. *The Origin and Nature of the Scottish Enlightenment*. Edinburgh: John Donald, 1982.

Cowan, Ian B. *Regional Aspects of the Scottish Reformation*. London: The Historical Association, 1978.

————. *The Scottish Covenanters, 1660–1688*. London: Victor Gollancz, 1976.

Cowing, Cedric B. *The Great Awakening and the American Revolution: Colonial Thought in the Eighteenth Century*. Chicago: Univ. of Chicago Press, 1971.

Daiches, David. *Scotland and the Union*. London: John Murray, 1977.

Dickson, R. J. *Ulster Emigration to Colonial America, 1718–1775*. London: Routledge & Kegan Paul, 1966.

Drummond, Andrew L. and Bulloch, James. *The Scottish Church, 1688–1843*. Edinburgh: Saint Andrew Press, 1973.

Drysdale, A. H. *History of the Presbyterians in England*. London: Publication Committee of the Presbyterian Church of England, 1889.

Fawcett, Arthur. *The Cambuslang Revival*. London: Banner of Truth Trust, 1971.

Ferguson, William. *Scotland's Relations with England: a Survey to 1707*. Edinburgh: John Donald, 1977.

Foster, Walter Roland. *The Church Before the Covenants: The Church of Scotland, 1596–1638*. Edinburgh: Scottish Academic Press, 1975.

Gaustad, Edwin Scott. *The Great Awakening in New England*. New York: Harper & Row, 1957.

Geertz, Clifford. "Religion as a Cultural System." In *Anthropological Approaches to the Study of Religion*, ed. Michael Banton. Edinburgh: Tavistock Publications, 1966. pp. 1–46.

Grubb, George. *An Ecclesiastical History of Scotland*. Edinburgh: Edmonston & Douglas, 1861.

Haire, J. L. M., ed. *Challenge and Conflict: Essays in Irish Presbyterian History and Doctrine*. Antrim, No. Ireland: W. & G. Baird., 1981.

Harvie, Christopher. *Scotland and Nationalism: Scottish Society and Politics, 1707–1977*. London: George Allen & Unwin, 1977.

Hatch, Nathan O. *The Sacred Cause of Liberty: Republican Thought and Millennium in Revolutionary New England*. New Haven: Yale Univ. Press, 1977.

Heimert, Alan. *Religion and the American Mind from the Great Awakening to the Revolution*. Cambridge: Harvard Univ. Press, 1966.

Heimert, Alan, and Miller, Perry, eds. *The Great Awakening*. New York: Bobbs-Merrill, 1967.

Henderson, G. D. *Religious Life in Seventeenth Century Scotland.* Cambridge: Cambridge Univ. Press, 1937.

Isaac, Rhys. "Dramatizing the Ideology of the Revolution: Popular Mobilization in Virginia, 1774 to 1776." *William and Mary Quarterly,* ser. 3, 33 (1976), 357–85.

————. "Evangelical Revolt: The Nature of the Baptists' Challenge to the Traditional Order in Virginia 1765 to 1775." *William and Mary Quarterly,* ser. 3, 31 (1974), 345–68.

————. "Religion and Authority: Problems of the Anglican Establishment in Virginia in the Era of the Great Awakening and the Parsons' Cause." *William and Mary Quarterly,* ser. 3, 30 (1973), 3–36.

————. *The Transformation of Virginia, 1740–1790.* Chapel Hill: Univ. of North Carolina Press, 1982.

Kenney, William Howland. "Alexander Garden and George Whitefield: The Significance of Revivalism in South Carolina, 1738–1741." *South Carolina Historical Magazine,* 71 (1970), 1–16.

————. "George Whitefield and Colonial Revivalism: The Social Sources of Charismatic Authority, 1737–1770." Diss., Univ. of Pennsylvania, 1966.

Killen, W. D. *History of Congregations of the Presbyterian Church in Ireland.* Belfast: Jas Cleeland, 1886.

Killinchy, or, the Days of Livingstone: A Tale of the Ulster Presbyterians. Belfast: Wm. McComb, 1839.

Klett, Guy Souillard. *Presbyterians in Colonial Pennsylvania.* Philadelphia: Univ. of Pennsylvania Press, 1937.

Kuklick, Bruce. *Churchmen and Philosophers: From Jonathan Edwards to John Dewey.* New Haven: Yale Univ. Press, 1985.

Landsman, Ned. "Revivalism and Nativism in the Middle Colonies: The Great Awakening and The Scots Community in East Jersey." *American Quarterly,* 34 (1982), 149–64.

————. *Scotland and Its First American Colony, 1683–1765.* Princeton: Princeton Univ. Press, 1985.

Leyburn, James G. *The Scotch-Irish.* Chapel Hill: Univ. of North Carolina Press, 1962.

Lodge, Martin Ellsworth. "The Great Awakening in the Middle Colonies." Diss., Univ. of California, Berkeley, 1964.

MacFarlan, D. *The Revivals of the Eighteenth Century, particularly at Cambuslang.* 1847; rpt. Wheaton, Ill.: Richard Owen Roberts, 1980.

Makey, Walter. *The Church of the Covenant, 1637–1651: Revolution and Social Change in Scotland.* Edinburgh: John Donald, 1979.

McClure, William, ed. *Extracts from the Writings of Ministers Connected with the First Presbyterian Church, Londonderry.* Londonderry, No. Ireland: James Macpherson, 1868.

McConnell, James. *Fasti of the Irish Presbyterian Church, 1613– 1840.* Revised by Samuel G. McConnell. Belfast: Presbyterian Historical Society, 1951.

McCoy, F. N. *Robert Baillie and the Second Scots Reformation.* Berkeley: Univ. of California Press, 1974.

McLoughlin, William G. *Modern Revivalism: Charles Grandison Finney to Billy Graham.* New York: Ronald Press Co., 1959. Chs. 1–3, pp. 3–165.

——. *Revivals, Awakenings and Reform: An Essay on Religion and Social Change in America, 1607–1977.* Chicago: Univ. of Chicago Press, 1978.

——. "The Role of Religion in the Revolution: Liberty of Conscience and Cultural Cohesion in the New Nation." In *Essays on the American Revolution.* Ed. Stephen G. Kurtz and James H. Hutson. New York: Norton, 1973.

Miller, Kerby A. *Emigrants and Exiles: Ireland and the Irish Exodus to North America.* New York: Oxford Univ. Press, 1985.

Miller, Perry. *From Colony to Province.* Cambridge: Harvard Univ. Press, 1956.

——. *Jonathan Edwards.* 1949; rpt. New York: Meridian Books, 1959.

——. "Jonathan Edwards and the Great Awakening." In his *Errand into the Wilderness.* New York: Harper & Row, 1956, pp. 153–66.

Moran, Gerald F. "Conditions of Religious Conversion in the First Society of Norwich, Connecticut, 1718–1755." *Journal of Social History,* 5 (1972), 331–43.

Morgan, Edmund S. *Visible Saints, The History of a Puritan Idea.* Ithaca: Cornell Univ. Press, 1963.

Murphy, Thomas. *The Presbytery of the Log College.* Philadelphia: Presbyterian Board of Publication, 1889.

Nash, Gary. *The Urban Crucible: Social Change, Political Consciousness, and the Origins of the American Revolution.* Cambridge: Harvard Univ. Press, 1979.

Nevin, Alfred. *Churches of the Valley.* Philadelphia: Joseph M. Wilson, 1852.

——. *History of the Presbytery of Philadelphia.* Philadelphia: W. S. Fortescue & Co., 1888.

Noll, Mark. "Ebenezer Devotion: Religion and Society in Revolutionary Connecticut." *Church History,* 45 (1976), 293–307.

Nybakken, Elizabeth, I. "New Light on the Old Side: Irish Influence on Colonial Presbyterianism." *Journal of American History,* 68 (1982), 813–32.

Perceval-Maxwell, M. *The Scottish Migration to Ulster in the Reign of James I.* London: Routledge & Kegan Paul, 1973.

Pilcher, George William. *Samuel Davies, Apostle of Dissent in Colonial Virginia.* Knoxville: Univ. of Tennessee Press, 1971.

Reid, H. M. B. *The Divinity Professors in the University of Glasgow, 1640–1903.* Glasgow: Maclehose Jackson & Co., 1923.

Reid, James Seaton. *History of the Presbyterian Church in Ireland.* 2nd Ed. 3 vols. London: Whittaker & Co., 1853.

Scott, David. *Annals and Statistics of the Original Secession Church.* Edinburgh: Andrew Elliot, 1886.

Scott, Hew. *Fasti Ecclesiae Scoticanae.* Edinburgh: Tweeddale Court, 1925. Volumes 1–7.

Stevenson, David. *Revolution and Counter-Revolution in Scotland, 1644–1651.* London: Royal Historical Society, 1977.

——. *The Scottish Revolution 1637–1644, the Triumph of the Covenanters.* Newton Abbott, Devon: David & Charles, 1973.

Stewart, David. *The Seceders in Ireland.* Belfast: Presbyterian Historical Society, 1950.

Stout, Harry S. "Religion, Communication, and the Ideological Origins of the American Revolution." *William and Mary Quarterly,* ser. 3, 34 (1977), 519–41.

Sweet, William Warren. *Religion in Colonial America.* 1942; rpt. New York: Cooper Square Publishers, 1965.

——. *The Story of Religion in America.* New York: Harper & Row, 1930.

Tracy, Patricia. *Jonathan Edwards, Pastor: Religion and Society in Eighteenth Century Northampton.* New York: Hill & Wang, 1980.

Trinterud, Leonard J. *A Bibliography of American Presbyterianism during the Colonial Period.* Philadelphia: Presbyterian Historical Society, 1968.

——. *The Forming of an American Tradition.* Philadelphia: Westminster Press, 1949.

Wallace, Anthony F. C. *Religion: An Anthropological View.* New York: Random House, 1966.

——. *The Death and Rebirth of the Seneca.* New York: Vintage Books, 1972.

Walsh, James. "The Great Awakening in the First Congregational Church of Woodbury, Connecticut." *William and Mary Quarterly,* ser. 3, 28 (1971), 543–62.

Webster, Richard. *A History of the Presbyterian Church in America, From its Origin until the Year 1760.* Philadelphia: Joseph M. Wilson, 1857.

Weir, A. J. *Letterkenny, Congregations, Ministers, and People, 1615–1960.* Photocopied booklet, located at Union Theological College, Belfast.

Williamson, Arthur H. *Scottish National Consciousness in the Age of James VI: The Apocalypse, the Union, and the Shaping of Scotland's Public Culture.* Edinburgh: John Donald, 1979.

Wolf, Stephanie Grauman. *Urban Village: Population, Community, Family Structure in Germantown, Pennsylvania, 1683–1800.* Princeton: Princeton Univ. Press, 1976.

Index